P9-EMQ-750

CO_2 Rising

CO_2 Rising

The World's Greatest Environmental Challenge

TYLER VOLK

The MIT Press
Cambridge, Massachusetts
London, England

For information on quantity discounts, email special_sales@mitpress.mit.edu.

Set in Bodoni, Galliard, and Futura by Graphic Composition, Inc., Bogart, Georgia, with InDesign CS2. Printed and bound in the United States of America.

Library of Congress Cataloging-in-Publication Data
Volk, Tyler.
CO2 rising : the world's greatest environmental challenge / Tyler Volk.
p. cm.
Includes bibliographical references and index.
ISBN 978-0-262-22083-5 (hardcover : alk. paper)
1. Atmospheric carbon dioxide—Environmental aspects. 2. Carbon cycle (Biogeochemistry) 3. Carbon dioxide. I. Title.
QC879.8.V65 2008
363.738'74—dc22

2008014027

10 9 8 7 6 5 4 3 2 1

for Marty, remarkable visionary

Contents

Preface

The most colossal environmental perturbation in human history is in the air. It is in every breath you take, in every wind, every cloud, each hurricane, and all skies everywhere, day and night. Ever-increasing concentrations of the heat-trapping greenhouse gas carbon dioxide (CO_2) are already disrupting the planetary energy balance, as is perhaps most visible in the melting of mountain glaciers and Arctic sea ice. The alterations to life caused by this increase are relatively mild so far, but they will increase in intensity in coming decades.

Consequences will sweep all portions of the globe. During this century and beyond, they will shape the climate, the diversity and distribution of living things, agriculture, water supplies, and the directions our technologies take. The rise in CO_2 will either help unite the emerging global society to counteract a shared alarm or divide it into competing camps.

When fossil fuels such as coal, oil, and natural gas are combusted, they release CO_2 as a waste by-product. That CO_2 is not an ordinary pollutant; it is an inherent result of the conversion of energy. It is ejected skyward, thereby letting loose carbon atoms that had been outside the biosphere system for millions of years during natural sequestration deep underground. Wafted worldwide, the waste carbon then infiltrates all parts of the surface biosphere system—the circulating webs of air, water, and soil, all of which include myriads of living creatures. In short, the newly created CO_2 enters what is known as the global carbon cycle. With loops within loops within loops, as in a Jackson Pollack painting, the global carbon cycle is one of the world's natural wonders. And unlike the Grand Canyon and other natural wonders, one need not travel to visit it. It's all around. We are embedded in the global carbon cycle, indeed privileged to live within it. And we are obliged to understand the profound changes that are unfolding.

The goal of this book is to set forth, in plain language and telling images, the essential facts about the dynamics of fossil-fuel-derived CO_2, what happens when it becomes part of the global carbon cycle, and how its creation is tied to humanity's material well-being. Great debates loom before us. What will be the concentration of CO_2 in the future? How will the new climate affect agriculture, ecosystems, and sea level? What energy systems will power civilization? Will collective decisions about energy policy be based on some degree of guesswork—informed though it might be—about the future dynamics of Earth's air, water, and life? Will we be able to continue global economic development yet protect ourselves against dangerous climate change? What are our odds of navigating safely into the future?

Definite answers to several of these difficult questions are either beyond the scope of this book or unknown. But I hope to provide a look at CO_2 and the carbon cycle that is both detailed and wide in scope, for the perturbation to the global carbon cycle lies at the core of the questions above. My aim is to not make the material too highly specialized. But neither will I hold back information that is essential to understanding the complexities of the challenge we face. I will tell you how the CO_2 increase was discovered at a laboratory on a Hawaiian mountain, how we now know from ice cores that the amount of CO_2 is already more than 33 percent greater than it has ever been in at least half a million years, and how the increase can be due to human activities even though it is a fact that soil bacteria generate an annual flow of CO_2 to the atmosphere that is nearly 8 times that generated by all our combustion of fossil fuels. You will see, for example, how there can be no net contribution to the CO_2 increase from the breathing of billions of humans.

Our bodies are made of carbon. So are trees. Carbon in the ocean is required by the floating hordes of tiny green plankton. Carbon is the structural core of carbohydrates, fats, proteins, DNA, and other biological molecules. The climate effects that we now read about and see almost daily in the media become much more salient when grounded in the material reality of how carbon shapes life and how the global carbon cycle links all organisms to one another and to the atmosphere.

Examining how wealth, energy use, and CO_2 emissions are linked, I will show how the emissions are tied to present and future trends in the global economy. From trends in the linkages, it is possible to work out why CO_2 will continue to increase, and at what rates, given what is known about the carbon cycle, and with a reasonable eye to the uncertainties. Reining in the increase will require further development of sources of energy that do not emit CO_2, such as carbon sequestration, solar, wind, nuclear, and a half dozen other potential answers to the need to deploy new sources of energy. There are big-picture issues that go beyond any debate, say, about a few wind turbines off the shore of Cape Cod.

Debates about the global environment, about climate warming, about fossil fuels, and about new energy systems have to face the issue of global equity on a per capita basis with respect to the CO_2 emissions, which spread globally no matter where they are emitted. For example, in the year 2050 will the United States' per capita CO_2 emissions still be more than 4 times the world average? Will the world even have an example of a complex, economically developed society with extremely low CO_2 emissions? The economic machine will probably roll on pretty much as it has been, at least for the global total, as developing countries produce more emissions. But what will the United States and other large per capita emitters do? The post-2050 odds on the gamble with Earth's climate will be more favorable if the countries with high per capita emissions start showing the way to a materially prosperous future with low emissions *right now*.

I begin the book by introducing the CO_2 molecule and its "greenhouse property" of blocking the very form of energy by which Earth cools itself. I also introduce the carbon cycle and—as a literary device—a carbon atom I call Dave.

Named after C. David Keeling, a distinguished carbon-cycle scientist, the carbon atom Dave serves (I hope) as a way of revealing the fascinating paths that actual carbon atoms take during their global circuits. I want to show you the carbon cycle from a carbon atom's point of view. Each chapter includes a vignette from the life of Dave the carbon atom. These vignettes relate to the substantive technical content that follows. In the first several chapters, for example, Dave's presence in an alcohol molecule in a glass of beer illustrates how a loop in the global carbon cycle works, his passage through a gas analyzer in the early 1960s leads to the topic of the discovery of the worldwide rise of CO_2, and his transit from the atmosphere into the ocean shows that circuits extend from plants to soils to air to water and back.

I was not able to introduce everything I wanted to by following a single atom of carbon. Dave, for example, entered the biosphere naturally—from the dissolution of a mineral in a limestone cliff during the last Ice Age. Some of the pathways in this book had to come from carbon atoms brought into the biosphere by the combustion of fossil fuels. Thus, I introduce several other atoms: Coalleen, Oiliver, Methaniel, and Icille. By tracing the stories of these atoms, we will look at the magnitudes of fluxes of different kinds of carbon in the biosphere, and the stability (or lack thereof) of the global carbon cycle in the past.

With the twin purposes of enhancing the enjoyment of our being alive as carbon-dependent organic beings and preparing the way for the later chapters, I planned the chapters in the first half of the book as relatively short primers on carbon fluxes that circulate among the great carbon-containing "bowls" of the biosphere. In the later chapters, I unfold the issues (already hinted at above) that are so challenging with respect to the future. These chapters too include episodes from the lives of my named carbon atoms—for instance, Oiliver and Methaniel are released from a burning stick used to cook a school lunch in Rwanda, and Dave passes through a wind turbine. But the material in the later chapters is denser, and the tone more pressing.

The book ends with Dave and the other named atoms making their separate exits from the biosphere many millennia from now. The fact of such long time scales is just one of the remarkable findings from the ongoing scientific investigation of the global carbon cycle, an endeavor in which I have been privileged to participate for more than two decades. Indeed, I hope that some of the process of science comes through in my discussions.

As you will see, I like graphs and charts. I am not trying to scare away readers who are not technically inclined; I am trying to facilitate a deep and personal understanding. I would be thrilled if you would look at my graphs and charts carefully, bring to them your own interpretations, and internalize them as pictures of how the great carbon cycle operates. It is my firm belief that we—as a global community—will have to understand the grand dynamics of the biosphere if we are to meet the challenges of the coming decades.

Acknowledgments

For direct help with points in the book or for conversations by voice or email about carbon, climate, and energy during the course of work, I thank Bob Berner, Sean Brandt, Long Cao, Ken Caldeira, Jenna Carlson, Eileen Crist, Nathan Currier, Tony Del Genio, Daniel Fink, Jon Foley, Joe Franceschi, Dale Jamieson, L. Danny Harvey, Skee Houghton, Joe LeDoux, Gregg Marland, Lauren Meriton, Lindsey Nelson, Brian O'Neill, Michael Raupach, Michael Rampino, Greg Rau, Denise Richter, Francesco Tubiello, and David Wolfe.

David Schwartzman, Mitch Thomashow, and Ken Volk read the entire manuscript and provided important comments. Susan Doll made technical remarks on chapter 7, Ralph Keeling on chapters 1–4. Special thanks go to Ralph for allowing me to grace the star carbon atom in this book with the name of his father. I will always remember far-ranging conversations with Dave Keeling (1928–2005), a scientist's scientist if there ever was one, and how he came up to me after my very first technical presentation to offer his encouragement. I would also like to acknowledge the support given early in my career by Taro Takahashi, another great carbon-cycle scientist.

Photos used in constructing figures were kindly supplied by Mark Siegal and Sasha Levy (2.2), Amelia Amon (3.1), Lyn Hughes (4.1), Susan Doll (7.1, 7.2), and Jeremy Young (10.1).

My bows to the professional staff at The MIT Press begin with my editor, Clay Morgan, who has been instrumental in developing a line of "biosphere" books and who boosted this project with his overall enthusiasm. I am grateful for the design skills of Yasuyo Iguchi, the administrative prompts of Meagan Stacey, and the copy editing.

In many ways, the book's conception was due to Janet Metcalfe, who cheered me on from its early days, and whose thoughts were invaluable to my own.

The book is dedicated to Marty Hoffert, my Ph.D. advisor and dear friend for nearly three decades. More than anyone, he taught me the tools for thinking globally by quantitative analysis and showed me the intellectual joys of such explorations. His courageous effort to wake up policy makers to the magnitude of the challenge ahead is an unending source of inspiration.

1

Introducing the CO_2 Molecule and Its Carbon Atom

My goal in writing this book is to provide essential information about what is surely the longest-term and most globally distributed environmental problem. I will aim for brevity, but I will lay out the essential inner workings of the global carbon cycle, concentrating on what I believe every global citizen should know.[1]

Carbon dioxide (CO_2) is the primary reason to be concerned about global warming and its consequences. The world appears locked in a certain direction. As CO_2 is increasing, so is civilization's dependence on fossil fuels, which creates ever-growing rates of CO_2 injection into the atmosphere. Climate effects will likely lead to serious disruptions in agriculture, coastal cities, human health, and the present web of life. Already, in the retreat of Arctic sea ice and mountain glaciers, and in shifts of the habitats of some species, we have warnings of change that will potentially be huge. But the issues are complicated. Not all the effects will be perceived everywhere as "negative." For instance, higher carbon dioxide, considered as an isolated factor, could enhance the growth of crops. However, the pertinent factors are interconnected, and they act as systems of feedback loops within the complex system of the biosphere.[2] I will unfold some of that complexity.

The CO_2 that goes into the air from combustion of fossil fuels does not stay there, but begins circulating throughout the biosphere. Furthermore, the fossil-fuel emissions are dwarfed by several natural inputs to the atmosphere, such as the constant releases of CO_2 from soil bacteria. Insofar as carbon dioxide is a perfectly natural gas that interacts with plants, algae, and the oceans—and indeed with everything in the biosphere—why doesn't nature just absorb the excess?

Understanding these and many more features that I feel are crucial for the drama's stage set is not particularly difficult, but the convoluted paths of carbon can stretch the imagination. Following those paths requires basic concepts, such as the presence of carbon in various chemical guises and magnitudes in distinct zones or regions. These zones usually are not separated geographically, as continents are—although they could be, depending on the focus of inquiry. More often they are distinguished by types of material and forms of carbon. Carbon-cycle scientists call the zones "pools." I will use that term, but I also like "bowls" (a more concrete metaphor). The four major bowls or pools of carbon in the system of the biosphere are the atmosphere, the soil, the ocean, and life taken as a collective bowl. The bowls are connected by a web of fluxes, which integrates the system and makes the dynamics of how carbon shifts around both tricky and exhilarating to follow. The scales of the fluxes and the bowls cross both space and time and vary from the minute to the gargantuan.

The human body is a complex system, too. When a doctor evaluates your condition and then discusses it with you, the doctor presumes that you know the basic facts about how your organs work: that the heart pumps blood, the lungs bring in oxygen, the stomach digests, the kidneys filter, the gonads make sex cells, the brain thinks, and so on. The same level of knowledge is not generally there about the organs of the planetary physiology. Thus, here I think of myself as an anatomy instructor. We are after the very anatomy of the biosphere, with its global metabolic pathways in and out and to and fro with the ocean, the soil, the air, and all living things. There are lots of people out there offering prescriptions for the ailing biosphere, from legitimate experts to snake oil salesmen. In reality, no one has the silver bullet solution yet. I believe that one quintessential component for hope is for citizens to become members of an adequately informed public able to digest and question the various debates on what to do, which will certainly rage over the coming years as increasingly serious prescriptions are put forward.

Doing nothing other than staying on the current course is equivalent to tossing dice for the future. Indeed, we have already entered an era of gambling with the global environment, and therefore it would be beneficial to at least understand the nature of the dice.

Here are the basics of why carbon dioxide is considered an environmental problem.

Earth Warmer: The CO_2 Molecule

Molecules are, of course, made up of atoms. In a CO_2 molecule, a single carbon atom is bonded to two atoms of oxygen on either side. Often the atoms in molecules are depicted as hard little balls, somewhat spread apart and linked by Tinkertoy-like struts, or as touching.[3] Neither of these simple ways of picturing atoms is strictly true. In reality, atom balls are electrical whirlwinds, and the bonds that link them are strong electrical affinities. The bonds that carbon atoms enter into with other atoms are formed by shared electrons that sweep and swirl back and forth between and among the atoms. Some renderings show molecules that look like partially melted ice cubes in surreal shapes. That helps us picture the atoms blended together, emphasizing the fact that

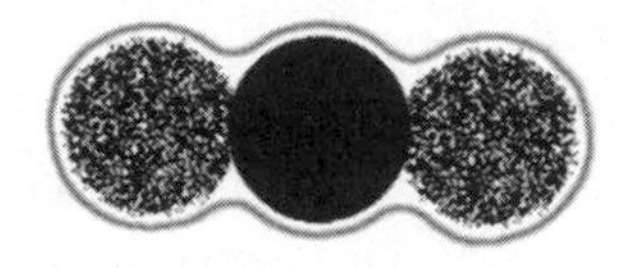

Figure 1.1 A CO_2 molecule, consisting of two oxygen atoms (shown in gray) and one carbon atom (shown in black).

the shared electrons form a truly new, higher-level entity: the molecule "above" the atoms. But no diagram can portray all the nuances of the reality, if only because the scale is below resolution by means of light (which is how we see objects that are close to our own scale).

In my diagrams, because I want to emphasize the individual atoms, I will draw simple balls, attached to one another and surrounded by an envelope to emphasize the reality of the molecule itself. Carbon atoms will always be black. Figure 1.1 shows the CO_2 molecule with its central atom of carbon.

The shared electrical bonds are not always exactly symmetrical. In the CO_2 molecule, the oxygen atoms are better at grabbing a portion of the shared electrical bonds than are the carbon atoms. That electrons have been pulled partially away from the carbon atom means that the carbon atom is in a relatively low energy state while in a CO_2 molecule. This has fundamental importance because we can burn fossil fuels, which before burning have their carbon atoms in higher energy states. The combustion that converts carbon from higher to lower energy states enables us to drive, to fly, to manufacture, to cook, to light our rooms, to heat our buildings, and to do many other things.

The story of fossil fuels and the rise in atmospheric CO_2 will unfold in the next few chapters, but first it is important to note how it is that this simple molecule possesses such powerful environmental effects.

CO_2 is a greenhouse gas. Molecules of greenhouse gases are capable of absorbing and re-emitting wavelengths of electromagnetic radiation that are in the infrared portion of the electromagnetic spectrum.

What is probably your most common experience with infrared waves occurs when you sit beside a fireplace or a campfire. Most of the air that is directly heated by the flames rushes straight up, so it is not this hot air itself that warms you; it is the copious flow of invisible infrared radiation, which is sent in all directions from the flames and the red embers.

Invisible to human eyes, infrared waves can be "seen" by sensors in the special facial hollows of pit vipers. This allows the vipers to detect temperature differences and thus to spot warmer prey (such as mice), even at night, against the background of the cooler soil. Military night-vision goggles take advantage of the difference in infrared emissions between the human body and cooler surroundings. Energy analysts who are hired to make buildings more efficient in winter often use cameras with special film or digital imaging to detect regions in a building's skin that are leaking heat.

What gives the CO_2 molecule its ability to absorb and (as a consequence of physics) re-radiate radiation in infrared wavelengths that are important to the atmosphere and the climate is its number of atoms: three.[4] "More than two" is the crucial concept here, because the other important infrared-capturing gases in the atmosphere also have three or more atoms in their molecules. When a gas has three or more atoms, it has modes of vibration inherent in its shape that can resonate with the frequencies of climate-affecting infrared waves. The matching enables the greenhouse molecules to intercept those waves and absorb their energy. Single atoms and two-atom molecules do not have those particular resonant modes. Just a few examples of other greenhouse gases make the point, if you count their atoms: water vapor (H_2O), methane (CH_4), nitrous oxide (N_2O), and ozone (O_3).

The two most abundant greenhouse gases are water vapor and carbon dioxide, with water vapor the most important in terms of total global infrared absorption. But it is CO_2 that is the environmental driver of Earth's climate. Why? Because the open bodies of available water that we call oceans allow water vapor to adjust as an effect of the primary heating caused by CO_2. The amount of water vapor, which will increase in response to the increasing concentrations of CO_2, is therefore considered a climatic feedback to the primary greenhouse effect of CO_2.[5]

Besides water vapor, three other gases in the atmosphere are more abundant than CO_2. Argon is 25 times as abundant on a molecule-to-molecule comparison, but it travels alone as one atom (a "noble" gas that snubs its nose at relationships) and has no greenhouse effect. The other two are oxygen (O_2), more than 500 times as abundant as CO_2, and nitrogen (N_2), more than 2,000 times as abundant. With only two atoms to each molecule, they are not greenhouse gases.

There may not be much CO_2 in the atmosphere, but its effects on the climate are powerful. Infrared rays, which Earth uses to cool itself to space, are exactly what CO_2 is good at absorbing and re-radiating. It does not matter that CO_2 molecules are present in tiny amounts. What matters is what CO_2 molecules do to the rays emitted from Earth's surface. This is analogous to the way a few drops of food coloring can tint an entire glass of clear water. The colored drops, dispersed everywhere in tiny concentrations, still affect the light enough to alter the entire look of the water. Small can be potent.

Figure 1.2 shows, in a highly simplified sketch, how the greenhouse effect works. Crucial to this picture is the concept of the planetary energy balance. Earth's surface is warmed by the sun. But the surface does

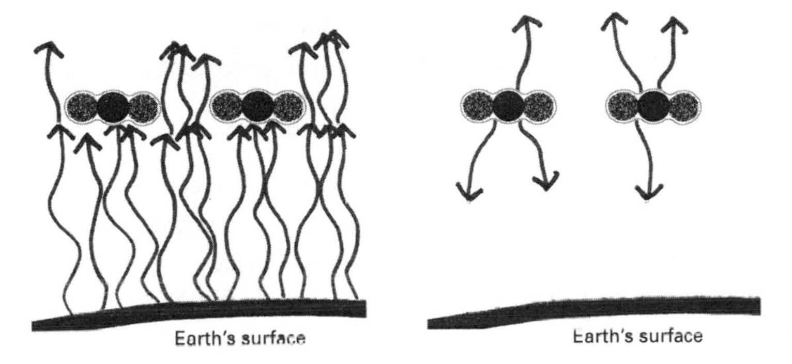

Figure 1.2 The greenhouse effect of CO_2. Left: To balance the energy absorbed from the sun (not shown), Earth's surface radiates rays of infrared waves to cool itself. Molecules of CO_2 in the atmosphere absorb some of the infrared rays. Right: Those same molecules, achieving their own micro-energy balances, then re-radiate new infrared rays in all directions. The re-radiated rays that happen to shoot downward exert an extra heating effect on the surface, which then will radiate even more infrared rays (not shown) to achieve a balance.

not get increasingly hot because the upper levels of the atmosphere radiate the cooling infrared wavelengths outward to the cold blackness of space. I have simplified the diagram to highlight the process of absorption and re-radiation of the infrared waves by the greenhouse gas.[6] In reality, the atmosphere becomes like a infinitely complex pinball machine for the rays, as they slowly work their way toward a net flow upward by a complex path, going up and then down, then up a bit more, then down not quite so far, then up some more, until eventually they get out free into space and the energy has left the biosphere. The greenhouse gases are like insulation in the walls of a home in winter. They block the exit of the heat. With the insulation, the heat stills gets out, but the inside temperature is higher as a result.

With the simplified picture and additional remarks, it logically follows that increasing the amount of any greenhouse gas results in higher temperatures (up to a saturation point of near-maximum effect of particular gases).

Without any greenhouse gases in the atmosphere, hypothetically assuming, say, that the CO_2 were reduced to zero and the water vapor as a feedback response dropped as well, calculations show that Earth's surface would be 60°F cooler (about 33°C cooler). With the average temperature of the surface now about 60°F (15°C), it is obvious that without its greenhouse gases Earth would be a frozen ball in space. The greenhouse gases are necessary for present-day life forms. They are why we have liquid water. Natural levels of CO_2 keep the planet from a permanent ice age. Yet rising levels of CO_2 threaten the future stability of the climate.

This, in a nutshell, is the physics of the environmental challenge we all must think about. As the CO_2 concentration rises, Earth is heading for a regime of greenhouse-gas levels and associated temperatures such as it has not experienced, as far as we know, in a very long geological time.

The Hero: The Carbon Atom

The carbon atom doesn't stay in the CO_2 molecule. Carbon dioxide is only one of the many molecules that contain carbon. Thus, to fully encompass what is going on with CO_2 we must expand the focus of understanding, gradually during the course of this book, to include the entire global, circuitous system of carbon in all its forms, in all places, and in many kinds of exchange. I have already hinted at this system, which is called the *carbon cycle*. The world travels of carbon atoms in the carbon cycle lead me to nominate a clear hero here: the carbon atom itself.

Carbon is the backbone of all biological molecules. Our bodies are carbon-burning machines fed by crops. Carbon fuels cooking fires in many developing nations. It also fuels most industrial energy systems. Civilization would collapse without the conversion of fossil forms of organic carbon into the waste gas CO_2. Biologically and technologically, we depend on carbon.

The same carbon atom can be in different kinds of molecules. A carbon atom, by breaking the electrical bonds to other atoms in one molecule and then forging new bonds with new neighboring atoms in new molecules, travels, in a sense, throughout the various "pools" of the biosphere. To make this clear, I will single out one carbon atom and give it a name.

The most famous scientist of the global carbon cycle was C. David Keeling, who dedicated his life to monitoring CO_2, starting half a century ago, and who discovered its rising levels in the atmosphere. In his honor, I give the name Dave to the main carbon atom we will follow.

Wherever Dave is right now, recently he was in an airborne CO_2 molecule. A few years before that, he was in the equatorial belt of the Pacific Ocean, in the type of molecule that holds (because of its total numbers in the vast ocean) the most carbon of any kind of molecule in biosphere: the bicarbonate ion, HCO_3^- (figure 1.3).

Furthermore, Dave recently was also in another kind of carbon-containing molecule that is widely abundant: a cellulose molecule in the trunk of a tree.

A cellulose molecule has an almost web-like appearance (figure 1.4). Other kinds of "organic" molecules have their carbon atoms in linear rows, such as the tadpole-like tails of lipids molecules in all cell

Figure 1.3 A molecule of bicarbonate ion, consisting of three oxygen atoms (shown in gray), one hydrogen atom (the smallest, also shown in gray), and one carbon atom (here labeled D for Dave, the primary atom whose path this book traces, which is often in a bicarbonate ion when in the ocean). The ion has a net single negative charge.

membranes. The cells and fluids of our bodies are filled with tens of thousands of kinds of proteins, among them the enzymes hemoglobin and insulin and the brain neurotransmitter dopamine. As molecules go, the enzymes are giants. Dave has been in all the types—in the lipid molecule of a membrane inside a little swimming copepod crustacean in the highly productive Bay of Bengal of Indian Ocean, in a humus molecule in the soil cast out an earthworm's rear end in the moist soil of Ireland, in the dopamine in the brain of a giant tortoise of the Galapagos Islands, and so on.

Clearly, Dave gets around. And he doesn't travel alone; he always takes up residence with other atoms, temporarily and for varying amounts of time that are not under his control. Sometimes his neighbors are the oxygen atoms in CO_2 and in the bicarbonate ion; sometimes his neighbors are other carbon atoms and hydrogen atoms, as in the cellulose and lipid molecules; sometimes he binds with nitrogen and sulfur atoms in protein molecules. Furthermore, his partners are different during these cycles. Every time he is in a CO_2 molecule, the

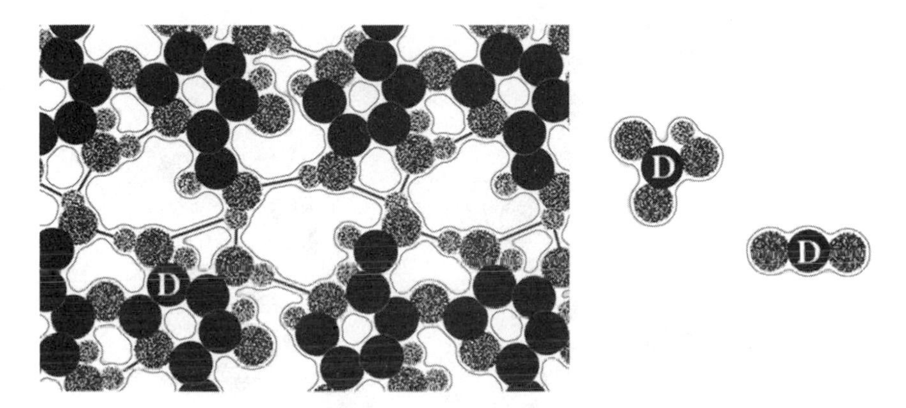

Figure 1.4 From left to right: Only a small portion of the huge network of the cellulose molecular system is shown, relative to the bicarbonate ion (middle) and CO_2 (far right). Carbon atoms, as is the custom in this book, are shown in black. Carbon atom Dave (D) has been in all these molecules and in others at various times; his location here in cellulose is arbitrary.

oxygen neighbors are different specific atoms of oxygen. Dave passes in and out of different molecular forms. And he can revisit those forms, creating sub-cycles within the global carbon cycle. Most trips among the pools for molecules traveling in the biosphere are not limited by one-way tickets.

The Sizes of Atoms and Molecules

Keep in mind when following the travels of Dave and the other carbon atoms that the scales are nearly inconceivably small.

Consider the CO_2 you exhale. It is a waste gas from your body's metabolism. And the concentration of CO_2 in your exhaled air is about 100 times that in the air that you inhale with each breath taken from the atmosphere. You, therefore, are a source of CO_2 to the atmosphere

Figure 1.5 "New" molecules of CO_2 are born from each human breath. These circulate in the atmosphere, and soon each leaf that grows incorporates some molecules from each breath.

(figure 1.5). The numbers work out to about 5×10^{20} molecules of CO_2 per exhalation.

When these new CO_2 molecules leave your mouth, they begin dispersing in the air. The entire atmosphere, from the North Pole to the South Pole, is stirred in about a year, so within that year the molecules you added from each exhalation are evenly distributed into the very air you will later breathe back in. To put the number 5×10^{20} in perspective, assume that you live in the mid latitudes of the northern hemisphere and that you exhaled at the end of the summer's growing season, say in

October. By the start of the next spring, the mixing in the atmosphere is essentially complete within the hemisphere. The green leaves that grow in the spring draw on CO_2 from the atmosphere as their source of carbon for the organic molecules they make, including cellulose and thousands of other kinds of molecules. Each and every leaf that grows will incorporate into its body a few dozen atoms of carbon that came from one particular exhalation you made during the previous fall.

2

From a Molecule in Beer to the Great Cycle of Carbon in the Biosphere

So where is Dave? We can't have him locked deep inside a rock, say in a bit of the organic content of a layer of dark gray shale half a mile under the surface of Arkansas, because then there would be no action to report—except, perhaps, to say "Come back in a few million years." Surely Dave should be somewhere in the active biosphere, the roiling, interconnected surface zone that includes the atmosphere, all waters and ice, all soils, and, crucially, all living things.[1] The biosphere—only 1/400 of Earth's radius—is somewhat like a nearly closed bottle in which we and most familiar living things thrive. And the stories of an atom of carbon within the biosphere could fill volumes.

When I was a child, a chemical company had the slogan "Better living through chemistry." With regard to organisms, the truth can be stated even more strongly: There is *only* living through chemistry.

Dave could be in CO_2, in marine bicarbonate, in plant cellulose, in worm castings, or in tortoise brain. He could be in any organism and in any of the tens of thousands of complex molecules made with the backbone of carbon-to-carbon links that give pre-med students taxonomic nightmares in the subject called organic chemistry. There are many possibilities for a story of Dave's shifty haunts in the interconnected chemical world in which almost any dream "body" becomes real: a myoglobin molecule inside a cell of a beating heart muscle, a structural cellulose molecule in the tree trunk outside the window, a light-sensing retinal molecule at the back of a soaring hawk's sharp eye, or, if he had come from a fossil fuel (which he did not), a styrene molecule in a sheet of plastic wrap that covers leftovers in your refrigerator, or the liquid in the gasoline tank of a car roaring down a new superhighway in India.

But for the moment we will bypass all these complex and large organic possible whereabouts for Dave and instead locate him in one of the simplest of all organic molecules. Its name is ethanol, its chemical formula is CH_3CH_2OH, and it is also known as ethyl alcohol, or, in more common parlance, simply alcohol. In view of its uses in the celebratory rites of most cultures throughout world history, ethanol rivals CO_2 in its impact on civilization.

Dave's current contribution is his presence in one of the two linked carbon atoms in that very molecule, perhaps in your local pub or your back yard. Dave is bonded to that other carbon atom and also to three atoms of hydrogen. Locked together, they float at the extreme sub-sub-microscopic scale in the brewed liquid of a freshly poured glass of beer (figure 2.1).

In addition to carbon atoms in the alcohol molecules of the beer, there are carbon atoms within CO_2 molecules inside the bubbles. Ethanol and CO_2, respectively, as the alcohol and the carbonation of beer, in addition to the flavorings from hops and other ingredients, create the essence of what makes beer enjoyable to many. In his current molecular incarnation, it was just a matter of chance, as we shall see, that Dave is in the alcohol and not the CO_2 of the bubbles. So how did Dave get into the beer? From the fact that beer is drunk fairly soon after bottling (unlike wine and whiskey, beer doesn't age well), we can quite accurately infer a good deal about Dave's recent past.

Figure 2.1 Carbon atom Dave in a molecule of ethyl alcohol (ethanol) in the liquid portion of beer within a drinking glass within the biosphere. Note also the numerous bubbles, which contain CO_2. The two carbon atoms in the ethanol molecule are shown in black, one large oxygen atom is shown in gray, and there are six small atoms of hydrogen.

How Dave Went from Air to Beer

A year ago, plus or minus a few months, Dave was in a CO_2 molecule in the atmosphere traveling across Germany above a field of barley. His two oxygen partners in that airborne ride, however, were not to be with him much longer.

It was summer. The air current swept low over the barley field and curled downward as part of a slow eddy that rubbed the air into the thicket of stalks of the barley plants. Within the dense growth, bits of air slowed to a standstill. A sub-visible micro-parcel of air containing Dave slipped into a long, arched green barley leaf through a tiny opening on its underside, one of hundreds of little doorways that allow gases to move in and out, to and from each leaf.

Inside the leaf, along tiny passageways somewhat like the branches inside our lungs, Dave (still in the CO_2 molecule) diffused first across the outer membrane of a tiny cell and then across a second internal membrane to enter a small football-shaped green chloroplast, where the action of photosynthesis takes place. Inside the chloroplast Dave was grabbed by the electrical attraction of a giant enzyme called Rubisco—a carbon-based, city-like molecule, an undulating, sticky giant blob of tens of thousands of atoms, and, as the crucial enzyme for all photosynthesis, the biosphere's most abundant enzyme.

Like a minister performing a marriage ceremony, Rubisco facilitated a bond between the CO_2 molecule and a seven-carbon molecule that had side groups of hydrogen and oxygen atoms. The marriage was doomed by the design of evolution, for quicker than overnight the new eight-carbon molecule divorced into two exactly equal four-carbon molecules. One of those molecules contained Dave.

Next Dave entered the Calvin cycle, the chloroplast's internal biochemical assembly circuit, in which countless molecules exactly like the one containing Dave were marched into a series of chemical reactions driven by the breaking apart of special energy-storage molecules. Imagine the processing stacks of an oil refinery that makes gasoline from crude oil. The ejection portion of the Calvin cycle converts the entering four-carbon molecules into six-carbon molecules of the simple sugar glucose. Having had some bonds with neighbors broken and new ones formed (which took energy), Dave emerged in a newly built molecule of glucose.

In the glucose molecule, Dave was more highly charged. As I noted earlier, when a carbon atom is in CO_2, the two oxygen atoms grab a lot of the shared electrical energy. But in glucose the carbon-to-carbon bonds are symmetric, the electrical bond is equally shared, and, as a result, the carbon atoms are more energized than when they are partnered with oxygen atoms. The barley leaf captured sunlight to forge this rearrangement of atoms, and therefore in the glucose molecule there is solar energy embodied in Dave and his new neighbors.

With embodied solar energy, Dave now had the potential to be burned with oxygen and, in terms of the overall chemical path, to go backward. Indeed, barley plants (like all plants) "burn" some glucose, speaking metabolically, back into CO_2. This pushes other glucose molecules into further biochemical assembly pathways that restructure the atoms of the molecules into even more complex shapes. All those tens of thousands of proteins (such as enzymes and other types of proteins), the lipid molecules of membranes, the cellulose molecules, and even the DNA and the RNA are made from simple raw materials, such as glucose, as well as nutrients such as nitrogen and phosphorus brought up by the roots. Thus, glucose is the source of both the carbon for further "Tinkertoy building" inside the plant cells and the energy needed by the cells.

On average, about half of the glucose produced in the Calvin cycle will be consumed by the plant. Had Dave been in that half, he would have gone fairly quickly right back into the atmosphere as CO_2, perhaps at night, for the plant doesn't need light for these so-called dark reactions. But Dave was in the half of glucose production that was reshaped into the more complex molecules that together create the organized magic of chemical reactions that constitute the networks of life. After several steps of such reshaping, and after what was for him a long trip down along a vein of the leaf, into the main stem, and then up into the plump barley seed in the plant's head, Dave ended up in a relatively permanent position as a carbon atom in an extended chain of what is considered a polymer of glucose, a molecule of a kind of starch.

About a month passed. Along came a mechanical harvester. Roaring through the now mature barley field, the harvester chopped down the stalk that contained Dave (still in a starch molecule in a barley grain), automatically threshed out the seed heads, and conveyed them into a giant hopper at the rear. From there Dave was vacuumed up into a storage bin. He sat there for several months until being shipped to the brewery, where the malt master took over the next stage of Dave's fate: to become malt.

The malt master set the controls for the right amounts of water, humidity, and air and forced the barley grains to sprout. They began to bud out into little seedlings, vital nubbins, as they might have become as seeds in a planted field after a rain. During the process of malting, the grain converted a portion of its inner storages of complex carbohydrates into more simple kinds of starches. At that point the malt master cranked up the temperature to put a sudden brake on further visible growth, but held the warmth at a level that allowed a natural enzyme inside the grain to continue turning the starch into a type of sugar molecule called maltose. We cannot know for certain all the details of Dave's conversion from starch to sugar, but it is likely that the bond to one of his carbon neighbors in the starch's chain of carbon atoms was severed and then he was reconnected into one of two linked loops of five carbon atoms — a transformation that has been orchestrated by humans for many thousands of years. (The ancient Egyptians made beer, even though they didn't know the molecular details.)

In the maltose, Dave was highly prized as a source of energy, especially to brewer's yeast. The malt master transferred the mash into the waiting molecular maws of living yeast cells (genus *Saccharomyces*). Pores in their membranes allow yeast cells to pull in maltose molecules and, without oxygen (for this process is anaerobic), split the maltose

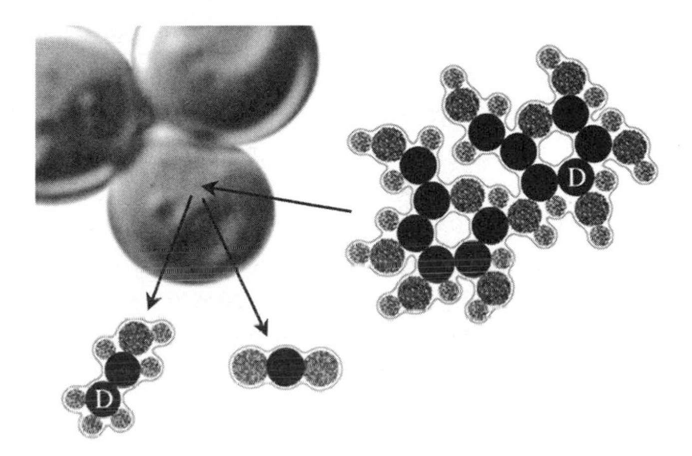

Figure 2.2 The relatively large maltose molecule (right) is taken in by a yeast cell and used for energy. Its waste products are alcohol (left) and CO_2 (middle). Here, carbon atom Dave comes in as maltose and exits the yeast in alcohol.

molecules off into molecules of alcohol and CO_2 (figure 2.2). From this bond-breaking, yeast cells derive energy to drive their own internal biochemical manufacturing of the diverse molecules specific to their bodies. In extracting energy from maltose, *Saccharomyces* is not as efficient as our aerobic (oxygen-using) body. Like most animals, we go the fungi one better: we can take all the carbon atoms in a sugar molecule and convert them into CO_2. But we can't make alcohol.

This was the point in Dave's travels, alluded to earlier, at which he might have gone into either a CO_2 molecule or an alcohol molecule. Much as the human body processes a piece of bread without caring which exit any one of the bread's atoms go out, it didn't matter to the yeast cell that processed Dave which path that particular carbon atom took. CO_2 and alcohol are simply waste products to a yeast cell as it grows and divides, feeds on maltose, and grows and divides again and again, excreting both of these wastes through its outer boundary membrane and into the fermenting solution of beer.

It was a year of changes for Dave. Here is a review of this most recent interval of his life in the biosphere, taking as a starting point the ethanol molecule in the chemical solution of beer and going backward: He had been in a sugary maltose molecule inside a yeast cell, and in beer malt. Before that, he had been in a starch molecule inside a barley grain. Before that, he probably had been in five or six other molecules in biochemical assembly lines inside the barley plant, which we can trace back all the way to the four-carbon molecule and then to the eight-carbon molecule he had been shunted into almost immediately upon wafting into the barley leaf as a CO_2 molecule that had been wind-blown and free in the atmosphere.

The Cyclic Pattern of Dave's Travels

Soon Dave will be back in the atmosphere. As ethanol, first in the beer drinker's stomach and then passed across the wall of his small intestine, Dave will enter the drinker's bloodstream. As the blood goes around and around the drinker's body and through his liver a number of times, eventually all the alcohol molecules are processed by the liver cells and treated as calorie-containing nutrients and also as toxins. Those liver cells then eject Dave and others into waste streams as bicarbonate ions (figure 1.3), which happens to be the most common form of carbon in the world ocean. When processing the maltose, the yeast had left some energy in the molecules of alcohol (from their inability to process all the carbon in the maltose to the low energy waste of CO_2, not from any altruism toward us). In a sense, our liver cells finish the job the yeast cells could not complete.

As the blood goes around and around the drinker's body, the waste bicarbonate ions in the bloodstream are converted into gaseous CO_2 in the lungs, which is then passed through membranes in the cells of the lungs and finally into the air within the branching network of bronchial passageways. Like water flowing as brooks merge into larger streams and then into ever larger rivers, the CO_2 moves along the lungs' hierarchical network of delicate tubes, into the mainline trachea, and up into the throat and then out the mouth or the nose into the atmosphere as part of an exhalation.

It might seem that opening and pouring a container of beer adds CO_2 to the atmosphere. The CO_2 in the beer's bubbles can rise up in the glass and enter the air when the bubbles hit the surface and burst, releasing their gases directly into the atmosphere. If the bubbles are drunk and go down the esophageal hatch, they can enter the bloodstream via the small intestine and then follow the path that Dave took, once he became bicarbonate in the blood, out of the lungs as CO_2 and into the air. And, as I have noted, the alcohol ends up as CO_2. Thus, in one way or another, the carbon in either of its two forms in the beer quickly becomes atmospheric CO_2. But while the beer can or bottle is sealed, that carbon is not in the air. Using the beer takes the carbon out of its storage and ultimately converts it to a gas, which would seem to have the effect of increasing the CO_2 content of Earth's atmosphere.

Yet because all the carbon in the beer originally came from plants (mostly barley, but also the hops used for flavoring), and because the plants procured all the carbon in their tens of thousands of kinds of organic molecules from CO_2 that had been in the atmosphere, there is zero net addition of CO_2 to the air from consuming the beer. To see the truth in this, we merely have to widen our scope to a period of a year or two. (Here I am neglecting the fossil-fuel energy that goes into the various stages of industrial agriculture; I am considering only the carbon atoms in the beer itself, all of which originally came from the atmosphere.)

We thus have a cycle. Dave left the atmosphere as CO_2 that went into the barley leaf, and after many molecular transformations he now will reenter the atmosphere as CO_2. By rejoining the air, Dave will complete one of the many possible circuits from a particular molecular form into several others and then back to the original form in the original pool (or bowl) of air. Dave's circuit is one of a diverse number of different round-trip loops that collectively create what is known as the global carbon cycle.

The pathways already described are just a few of the essentially infinite ways of completing subcycles in the global carbon cycle. There are more than 250,000 species of plants that Dave could have entered in the first place, and only in a few of them would he have ended up in beer as part of the subcycle.

There is a structure revealed or at least hinted at here, and that structure is crucial for thinking about the planet. To borrow a term from Hinduism, many of the paths of transformation of a carbon atom are portions of cycles of reincarnation.

In a fundamental sense, the analysis can be generalized to apply to all the CO_2 expelled by human breaths, not just the exhaled CO_2 derived from metabolically processed beer. The cells of our body convert about 80 percent of the carbon in the organic molecules of our various foods and drinks into CO_2 that we exhale out through our nostrils and mouths. Whether we ate animals, fungi, or plants as the source of the organic molecules, all that ingested carbon ultimately came from plants that took it from the atmosphere during photosynthesis. So it does not matter how many humans are breathing. The carbon dioxide in the global human total of 100 billion exhalations per minute is merely returning to the atmosphere carbon that was recently snatched from the atmosphere by crops. Thus, the human population, considered as a purely biological herd of nearly 7 billion organic beings, cannot cause a rise in atmospheric CO_2 from its exhalations.

The remaining 20 percent of our ingested organic carbon exits the body as urea in urine and as various molecules in feces. Those are highly valuable as sources of carbon and energy to the bacteria that thrive in urban waste-processing facilities (cultivated by sewage engineers as very cheap labor), or to the bacteria that live in the underground distribution tiles of rural septic fields, or to the bacteria that live in the soil of woodlands. The bacteria then complete the cycle by turning the waste of the human animal back into CO_2.

With the release of human-derived CO_2 that comes not from food or drink but from the combustion of fossil fuels, the story is quite different, as we soon will see. But right now, let us look back a bit further into Dave's past. Let us go back to a special day more than 40 years ago.

3

The Worldwide Increase of CO_2

There are two reasons I have singled out Dave. First, in the beer in an alcohol molecule, he served as a simple yet illustrative example of a closed loop. Tracing him in time both back to when he was in atmospheric CO_2 and forward to when he will be in atmospheric CO_2 illustrated one small sub-cycle within the global carbon cycle of Earth's biosphere. Second, an event in Dave's past warrants that we honor him with a spot in the high echelon of carbon atoms. That event took place in the early 1960s.

Being singled out is apparently no easy task for a carbon atom, given that in the atmosphere alone there are 42×10^{39} molecules of carbon dioxide. This number is beyond any reasonable reach of imagination. Yet the very existence of such multitudes—in other words, the infinitesimalness of atoms in the immensity of the biosphere—allows me to point out one atom that is now in a glass of beer and that achieved a kind of collective notoriety more than 40 years ago. I can't pluck out for you the exact atom that is Dave. But just as I can point to any fresh leaf in springtime and assure you that some carbon atoms in that leaf came from every single exhalation that you made last year, I can assure you that there is somewhere such a Dave.

Measuring Carbon Dioxide

Human senses are not tuned for detecting CO_2, which is why we call it an odorless, colorless gas. If we did detect it, by definition it would have smell or color, because our senses would give it qualities. In our evolutionary past we did not have any need to develop sense organs for this gas. But, just as some animals can detect infrared waves that we cannot, some of them do sense CO_2 levels outside their bodies. The mosquito, for one, uses concentrations of airborne CO_2 to help it locate a living mammal, such as a moose or a human, whose breaths continually disperse a cloud of elevated levels of CO_2 around its body.

We do react to CO_2 to the extent that high concentrations cause us to suffocate. Even then, we could not discern whether we were asphyxiating from high CO_2 or low O_2. As already noted, however, our body's ability to regulate the concentration of substances in the blood does entail a kind of monitoring of the internal amount of CO_2 (bicarbonate ion, in fact), which drives the pulmonary exhalations that expel this toxic waste by-product from the metabolism of each and every one of our 10 trillion body cells, from cranial nerves to muscle cells in the big toes.

To measure CO_2 in the air, we turn to analytical techniques developed by chemists. One popular piece of equipment makes use of the gas's greenhouse properties—that is, its ability to absorb and reradiate infrared waves. So it was that in the early 1960s, during measurements by the atmospheric chemist David Keeling, Dave the carbon atom found himself in a small parcel of air passing through a metal tube with infrared-transmitting windows inside an infrared gas analyzer.

In 1800 the German-born English physicist William Herschel had inferred the existence of infrared rays from his investigations of how a thermometer was affected by the various bands of color from sunlight passed through a prism. The prism created a "rainbow" spectrum that spread across Herschel's lab table. He didn't know it at the time, but the prism also split off the infrared waves into their own zone on the table, just beyond the visible red band of the spectrum. When Herschel laid the thermometer beside the band of red, just outside the visible spectrum, he found the highest temperature of all. He concluded that there must be wavelengths that had energy that he could not see, but which the sensitive fluid in the thermometer could absorb and thereby reveal.

Most of the energy emitted by the sun is in the portion of the electromagnetic spectrum that we call visible light. But the sun's rays contain energy in the invisible wavelengths, too, including infrared, ultraviolet, and radio waves. Earth, on the other hand, radiates primarily in the infrared range (and not in the visible range at all; it is much too cold to glow). At ordinary Earth temperatures, all the things around us (including human bodies) give off infrared radiation as their primary type of emitted electromagnetic waves. And, as I have noted, infrared waves are crucial for understanding the greenhouse effect, because Earth establishes its average planetary temperature by reaching a balance between the solar energy absorbed and the infrared radiation sent into space. The presence or absence of greenhouse gases that absorb and reradiate infrared rays affects how that balance is achieved, and therefore these gases are major determinants of global surface temperature.

The interaction between greenhouse gases and infrared rays has been put to practical use in the infrared gas analyzer, which measures the absorption of infrared rays to compute the concentrations of carbon dioxide and other gases.

Where Dave Was Analyzed and What Was Found

In the late 1950s, David Keeling, with a newly minted Ph.D. in atmospheric chemistry, had developed ways of measuring CO_2 more precisely than ever before. He understood that to measure CO_2 continuously could be of vital scientific importance, for it was known that the combustion of fossil fuels was releasing the gas into the air. But no one knew what happened next to those emissions. Measurements had been started at several locations as part of the International Geophysical Year. In particular, Keeling obtained funding to set up two analytical laboratories: one at the meteorological observatory at the South Pole and one on the north flank of the huge shield volcano called Mauna Loa on the Big Island of Hawaii (where the National Weather Service already had an atmospheric monitoring station). Keeling's goal was to initiate continuous and precise monitoring of the atmosphere's CO_2 content.

Keeling argued that the Mauna Loa observatory—more than 10,000 feet above sea level, and near the geographical middle of the Pacific Ocean—was an ideal place to collect samples year-round with minimal influence from the continents' urban pollution or their vast forests. I have been that high on Mauna Loa's neighboring volcanic peak, Mauna Kea, whose top now sports an array of deep-space telescopes. At such altitude, the air is cold, the sky clear and bright. To a visitor coming up from sea level just for a day, the height is dizzying (figure 3.1). The landscape is as barren as that of Mars. Hawaii's tropical lush greenery seems like another world.

More than a century before the establishment of Keeling's lab, it was well known that without CO_2 in the atmosphere all plant life on Earth would die, thus dooming all animals and even the soil's detrital food webs of fungi and bacteria (which require fresh litter falling down from above). A second reason why CO_2 is essential to life was also known:

Figure 3.1 From near the top of Mauna Kea, on the Big Island of Hawaii, the author points toward the National Oceanic and Atmospheric Administration's observatory on the flank of Mauna Loa, where CO_2 and many other gases are monitored.

the greenhouse effect from CO_2 and water vapor feedback keeps Earth's surface about 60°F (about 33°C) warmer that it would be otherwise. But the precise amounts of CO_2 in the atmosphere were not known, and many scientists believed the gas to vary from place to place much more than Keeling would eventually show to be the case.

By the early 1960s, Keeling's lab had been in operation for several years. Dave the carbon atom, in his molecular incarnation within an airborne molecule of carbon dioxide, had been free in the air for about two years. Wafted along on a northeasterly wind, and bobbing slightly around 11,000 feet, Dave passed high above the beaches and jungles along the Big Island's eastern shoreline. Ahead, to the west, was Mauna Loa.

Dave traversed several craggy outcrops on the mountain, descended about 1,000 feet, and entered a small air intake on the side of an isolated building. He was channeled into the lab and then into a large metal box that contained a tube for air samples. The air parcel containing Dave was about to be bombarded with infrared rays emanating from a heating element within the gas analyzer and sent into the tube through a window.

The CO_2 molecule to which Dave belonged absorbed many infrared rays during its brief stay in the glass tube. After each absorption, the vibrational energy of the molecule jumped to a higher state. Like an egg balanced on one end, this state was inherently unstable. The excess vibrational energy state was quickly quenched within the molecule, which as a result became warmer; then it released a new infrared wave in a random direction as the molecule returned to its vibrational "ground state." The capture of the original rays and the reemission of new rays in all directions had the net effect of reducing the number of original rays that made it across the tube to an exit window.

The more abundant molecules of nitrogen and oxygen had no effect on the infrared rays in the air sample. Thus, by using thermal detectors that were (in an expanded sense) the sophisticated descendants of Herschel's simple experiment, the device could record what fraction of a very specific wavelength of infrared ray made it through the sample of air—a wavelength whose blockage would reveal the amount of CO_2 gas in the air sample. The specific infrared wavelength in the gas analyzer was not in the range of wavelengths that are relevant to Earth's greenhouse effect. Nonetheless, in essence the properties of carbon dioxide that cause it to be a greenhouse gas were being used to identify the amounts of the gas, to better compute the potential future greenhouse effect. The higher the amount of CO_2 in the tube, the more the rays passing through the tube were reduced, allowing the atmospheric scientists to read off the concentration of the CO_2.

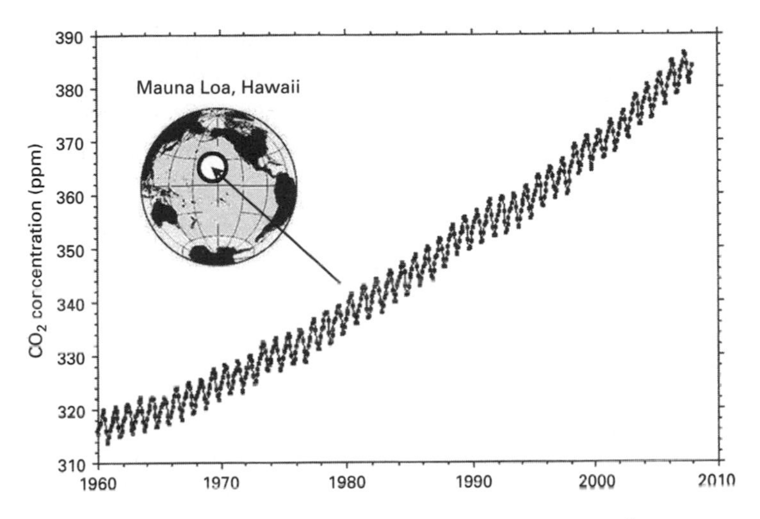

Figure 3.2 The CO_2 concentration at NOAA's Mauna Loa observatory. Each dot represents a month. The peaks are in May.

The data from Dave's air parcel and from many other air parcels were published in several early papers by Keeling, and those papers became instant classics in the legacy of science. They showed precise results for atmospheric CO_2 measured at Mauna Loa.

There are now 50 years of data[1] from the Mauna Loa observatory. The graph reproduced here as figure 3.2 is probably the most famous graph of the last 100 years, and its fame is destined to continue for many decades if not for centuries.

One big surprise was that the CO_2 varies up and down systematically. If you count the cycles in the plot, you can easily verify that they are annual. The data follow the seasons: up during fall, winter, and early spring, then down during late spring and summer.

But the major result, which was clear even from the first few years of the Mauna Loa data, was the steady upward march of the average level of carbon dioxide. The late-spring peak of any given year is larger than the peak of the previous year. The autumn low of each year is not quite as low as the low of the year before. The yearly averages keep going up. The Mauna Loa data opened the eyes of humanity to the fact of rising CO_2.

As I noted earlier, today there are 42×10^{39} CO_2 molecules in the atmosphere. In the early 1960s there were only 35×10^{39}. That is, there are 7×10^{39} more CO_2 molecules in the atmosphere now than there were in the early 1960s. We can put that number in perspective by using better units.

The most convenient unit for talking about the air's carbon dioxide concentration is parts per million (abbreviated ppm). Today (in 2008) the concentration of CO_2 is about 385 ppm. This means that if you were to grab a handful of air (and remove the water vapor, which varies with the humidity) only 385 of every million molecules of air would be molecules of carbon dioxide. That is 0.038 percent. To repeat what I trust you will deem worthy of repetition, this seemingly small amount of CO_2 can be powerful, like a drop of food coloring in a glass of water, because CO_2 is the primary greenhouse gas whose changing level sets Earth's climate (whereas the more abundant greenhouse gas, water vapor, is considered a feedback that adjusts up and down with CO_2).

From the early 1960s to today, the concentration of CO_2 has increased about 20 percent. Whatever the countries of the world decide to do, the numbers will continue to climb for decades to come, because large-scale action to curb the rise will almost certainly be slow. Later in the book we will see just how much the concentration is expected to rise. But I have no doubt that the Mauna Loa graph is destined to be known by every human alive today, and even by those not yet born. The story in this graph is destined to be in the news media every year for the rest of our lives.

But the elevation of this graph's fame because of its environmental import depends on the data's being representative of the atmosphere as a whole. We are talking about a gas-measurement lab on a volcano, after all. Often the first thing someone asks when shown the Mauna Loa data and told the location of the lab is "What about the volcano?" Could the volcano be skewing the data? Volcanoes do give off carbon dioxide, among other gases.

It would be easy to say that the atmospheric scientists were sure that the volcano would not be problematic, otherwise they wouldn't have chosen that site. Perhaps they could simply discard data suspected of being influenced by an eruption. Indeed, emanations of CO_2 from the top of Mauna Loa do get detected as occasional blips in the data, which can easily be distinguished like power glitches in a radio signal.[2] Fortunately, Mauna Loa is a shield volcano that is currently not erupting at its top but rather along fissures on the other side from the laboratory and far below, at an altitude where lush vegetation can inhabit cracks in old eruption beds and where people go every day and night hoping to see red-hot magma steam into the sea. How handy that the low-elevation fissures are erupting in a place which the United States designated a national park! The only downside is that on some days visitors are turned away because of the sulfur gases.

But let us approach the entire situation with what the Zen tradition would call beginner's mind, or with what those in the tradition of science would express as "I'd like to see for myself, please." Let us look at the results from the lab at the South Pole.

Carbon Dioxide at the South Pole

Fortunately for us, David Keeling had the same idea. The South Pole was even more remote from cars and vegetation than Mauna Loa, and the U.S. government was already operating a meteorological station. Keeling received funding to begin monitoring CO_2 there.

The South Pole has 6 months of light, 6 months of dark, and sunrises and sunsets in between that last a week each. From the station's website it appears that the scientists there get a little wild, perhaps to gain release from the unrelenting cold, the extreme seasonality of light and dark, and the isolation. Men sometimes don horned Viking hats, and women party outdoors in clothes too scanty for the cold, to maintain their humor during concentrated technical work. Inside, an infrared analyzer grabs air samples piped in from outside and measures the CO_2 concentrations.

The monitoring of CO_2 at the South Pole and at Mauna Loa provide the two longest-running historical records. Figure 3.3 shows monthly values at both sites.

One immediate point to notice is that the South Pole data show very little seasonal oscillation. They do show small yearly wriggles, but these are small in comparison with the annual rises and falls at Mauna Loa. This contrast in the behavior of the annual cycle, which will be shown to be primarily a difference between the northern and southern hemispheres, provides a deep insight into how vegetation interacts with the atmosphere and into how powerful the forces of nature are relative to those of humans.

A second point, and certainly the main one to be taken from figure 3.3, is that the South Pole data show nearly the same general rising trend as the Mauna Loa data. Although these two sites are far apart and far from the main hives of industry, the rise in CO_2 has been about the same.

In the late 1950s, the CO_2 concentration was about 315 ppm at both sites. Fifty years later, the CO_2 concentration at Mauna Loa was about 385 ppm, while that at the South Pole was 382 ppm. Thus, over the duration of the record, the concentration increased by 22 percent at

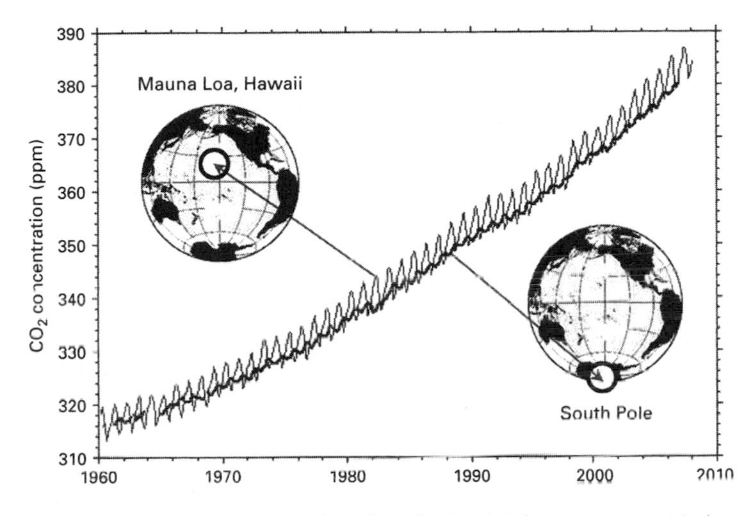

Figure 3.3 Atmospheric CO_2 data from the South Pole Observatory (darker line), compared with data from Mauna Loa (oscillating lighter line, with monthly dots removed for clarity).

Mauna Loa and by 21 percent at the South Pole. Mauna Loa appears to be slowly surpassing the South Pole, as can be seen from the graph. The reason is that more CO_2 is released from fossil-fuel combustion in the north than in the south. But the main conclusion is not the small difference in rates of growth over a half-century for the two sites; rather, it is that, generalizing from these two sites, the rise in CO_2 appears to be truly global.

Keeling's published reminiscences help us appreciate what these measurements demonstrate.[3] He tells how one of the most famous Earth chemists at the time was Roger Revelle. In the film *An Inconvenient Truth*, Al Gore cites Revelle as the professor who inspired him to get concerned and technically informed about the rise in CO_2. According to Keeling, Revelle, a major figure in geochemistry, "held to the prevailing belief that the CO_2 concentration in air was spatially variable and that therefore sampling must be widespread to establish a reliable global average." Specifically, Revelle called for aircraft to be used in a sampling program.

Furthermore, Keeling told how Scandinavian scientists presented their own data at an international meeting in Finland he attended in 1960. The Scandinavian program, begun years earlier by the meteorologist Carl-Gustav Rossby, had yielded results that showed atmospheric CO_2 varying from 150 to 450 ppm, in line with Revelle's belief that the concentration of the gas changed greatly with location and would thus require a large array of measurements just to get a simple global average. The Scandinavians even hoped to use CO_2 as a tracer to monitor the movement of air masses that differed in their CO_2 concentrations. Keeling's precise measurements, which he presented at the meeting, led to the demise of the Scandinavian program (which, it turned out, was subject to technical errors) and to the demise of the hypothesis of huge geographic variability in atmospheric CO_2. In truth, as figure 3.3 shows, either the Mauna Loa data or the South Pole data alone gave a global average good enough to tell us that CO_2 was rising and at what rate.

The CO_2 data counter many of our intuitions about the atmosphere. Most of us probably have an intuition about the dynamics of air pollutants derived from our visual or olfactory experiences of car exhausts or fumes from industrial smokestacks. Black, gray, yellow, or white clouds of pollutant particles disperse outward from their sources but are most hazardous in areas near where they are emitted. The worst place to be standing when you are hit by exhaust from a truck is right behind the tailpipe. To be sure, the chemical pollutants that form acid rain can cross state and even national boundaries. But these, too, are limited in extent, because the ions of sulfur and nitrogen soon leave the atmosphere in raindrops. The radioactivity spewed into the air by the

1986 nuclear accident at Chernobyl had its worst effects on the regions just downwind from the accident site. Even natural storm systems dissipate as they travel.

Carbon dioxide is different. It is not a particle that falls out or an ion that rains out in amounts substantial enough to cleanse the atmosphere after a few storms. It is a chemically very inert gas. Its molecules travel in the air until pulled into the ocean or into land plants. And, as will be described later, these great pools of carbon also give the CO_2 back. But such exchanges happen slowly relative to the fast mixing of the atmosphere. Some of the early scientists involved in the carbon cycle, though they knew the atmosphere to be stirred by the winds, had incorrect intuitions about how thoroughly the giant ocean of air from which we draw our breaths is blended. The fact that air samples taken at Mauna Loa and at the South Pole show the same amount of rise over 50 years demonstrates the essential unity of the atmosphere. The circulations of the tropical easterlies and the middle-latitude westerlies, the clockwise and counterclockwise gyres that swirl around the great high- and low-pressure systems of both hemispheres, and the tremendous uplifts and falls of the bellows called the Hadley cells in the low latitudes thoroughly mix the atmosphere in about a year, all the way from the high northern hemisphere to the South Pole. That is why in about a year CO_2 from your exhalation is found in all growing leaves in both hemispheres.

Alaska, the Tropical South Pacific, and Other Sites Confirm the Finding

David Keeling's work did not end the need to monitor CO_2 across different sites. For one thing, scientists still wanted to track that rise at a number of places, because the differences were providing them with

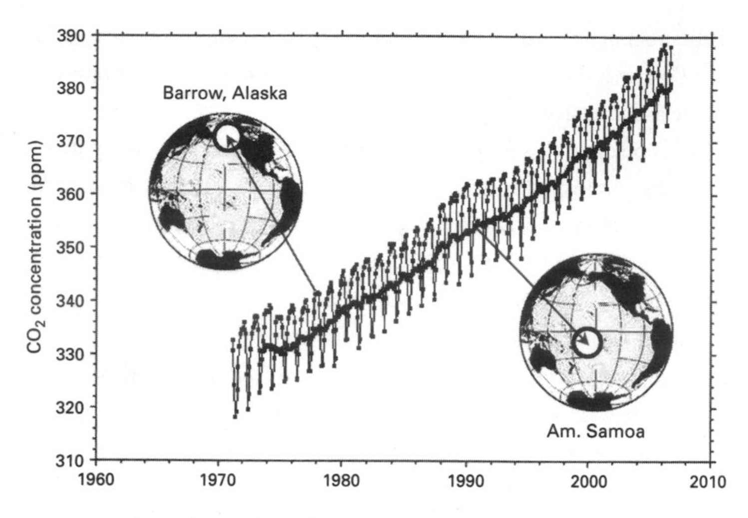

Figure 3.4 Data from additional sites that have relatively long time series for CO_2: Barrow, Alaska (71°N) (light line with monthly dots and large seasonal oscillations) and American Samoa (14°S) (dark line with almost no seasonal oscillations).

crucial information about the subcycles of the global carbon cycle. So, in a sense, Revelle was right. The seasonal oscillations, strong at Mauna Loa but weak at the South Pole, should certainly be examined at more locations. Keeling quickly pegged the seasonal swings as coming from vegetation.

In the mid 1970s, the National Oceanographic and Atmospheric Administration began monitoring above the Arctic Circle at Barrow, Alaska, on the coast of the Arctic Ocean. Later, monitoring was initiated at American Samoa, a tropical South Pacific island. Figure 3.4 presents data from these two sites.

Data from Alaska and Samoa fit right in with the trend from Mauna Loa and the South Pole, where monitoring was begun nearly 20 years earlier. We are witnessing a global phenomenon. CO_2 is rising everywhere, and at about the same rate.

The annual oscillations continue to generate interest. The seasonal cycle at Barrow is even larger than that at Mauna Loa—indeed, more than twice as large. And there is almost no seasonal cycle at Samoa, a pattern that turns out to be indicative of data from across the southern hemisphere.

What about comparing a site in the Atlantic Ocean and a site in the middle of Europe? Air blowing east across the Atlantic should then pick up CO_2 emissions from all the industrialized European nations. Figure 3.5 shows data from the Portuguese Azores, in the North Atlantic, alongside data from Mount Cimone, a meteorological observatory in Italy. Again, the rate of the increase is the same. Yes, the level of carbon dioxide is higher in the center of any major city. But the greenhouse effect is a large-scale property of the atmosphere. And the mixing of the atmosphere spreads the emissions from any particular nation all around the world.

Figure 3.5 also shows data from Ascension Island in the South Atlantic and from New Zealand. These sites in the southern hemisphere exhibit the same pattern seen at American Samoa and at the South Pole: a very small seasonal cycle relative to the dramatic seasonal cycle at sites in the northern hemisphere. These patterns perplex carbon-cycle scientists who are trying to figure out how much CO_2 goes into a forest or into the ocean's surface in a particular region, or how much is exiting from that region, but on average the CO_2 content of the air is the same everywhere on the planet.

Can we be absolutely sure that the rise is due to human releases of CO_2? Perhaps there is some other biosphere-scale phenomenon happening. This is an important question, and I will spend time on it in other chapters. But for now, the data are clear. If we assume that

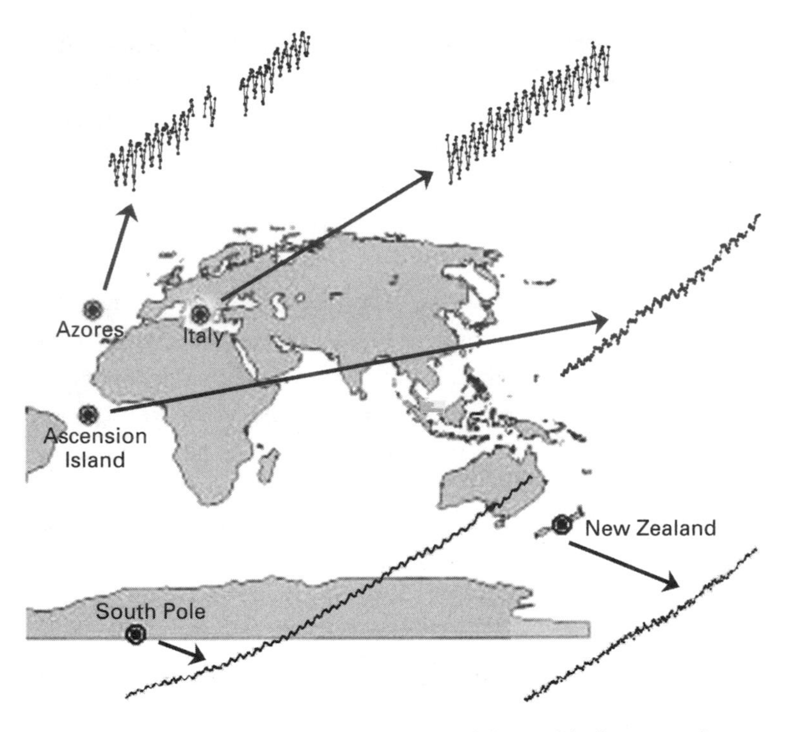

Figure 3.5 CO_2 data from more sites around the world. The Azores have some data missing. The South Pole is the longest-running times series here.

the rise is attributable to human activities (and please go ahead and assume that), then the effects of those activities have spread equally everywhere. Perhaps it is surprising to imagine how a planet this big can have an atmosphere so well mixed, but we witness it in the data. No matter how difficult it might be to start thinking on a truly biosphere-size scale, that is what all the data tell us we must do.

The atmosphere is singular. The CO_2 released from the combustion of fossil fuels from any country does not pollute that country's air any more or any less than it pollutes the air in all other countries. No one country can complain that it is special because it is downwind from any other particular country. Of course, some countries can complain that they are releasing only a little carbon dioxide while others are releasing the bulk of global emissions. But such complaints have nothing to do with geography. The atmosphere's mixing makes contributions to CO_2 from any place affect all other places equally—all countries, forests, polar regions, and oceans.

4

Fossil-Fuel Carbon Atoms Join the Biosphere

When faced with things that are too big to sense, we comprehend them by adding knowledge to the experience. The first appearance of a shining star in a darkening evening sky can take you out into the universe if you augment what you see with the twin facts that the star is merely one of the closest of the galaxy's 200 billion stars and that its light began traveling decades ago. The smell of gasoline going into a car's tank during a refueling stop, when combined with the datum that each day nearly a billion gallons of crude oil are refined and used in the United States, can allow our imagination to ripple outward into the vast global network of energy procurement and politics.

The biosphere is like the universe and geopolitics. Local bits of experience can be savored at a new level by imaginatively bringing in knowledge. The sensory response to a meadow full of autumn flowers pulsed by slow shadows cast from flotillas of cumulus clouds can be boosted outward in a big leap to include the oceans, which supply most of the water for the clouds and the growth of flowers. The main message from the sciences of air, soil, and ocean is that the biosphere is a gigantic system of mixing, an Earth-scale stew in constant circulation, in which the smallest bit of local color is linked to the entire spectrum, both seen and unseen.

To an ordinary observer on the local scale, the biosphere seems vertically thick, because it stretches from the ocean's depths up to the atmosphere's upper boundary, fading gradually into space far above the reach of high-flying intercontinental jets. But relative to the entire sphere of the planet, this surface layer is as thin as the shiny skin is to an apple—only about 1/400 of the radius. The biosphere's real immensity is in its horizontal spread. It wraps the globe, embracing all latitudes and longitudes of ocean, air, and soil and all creatures. The biosphere is not a protective skin like that of an apple. It is roiling with energy, more like the colors undulating on a soap bubble; a spherical shell; a dynamic cauldron within which both physical, chemical, and biological forces boil and blend together so well that Antarctic penguins and Arctic polar bears are only a year apart from inhaling gases exhaled by one another.

The biosphere is stirred not only by winds that mix the air, but also by waves and currents that mix the oceans and by chemical reactions and worms that mix the soils. The oceans and the soils mix more slowly than the air. The ocean (considered as one) takes about a thousand years to turn over from its rapidly stirred surface down to its pitch black and everywhere cold depths. Litter at the surface layer of the soil is churned by worms and millipedes in just a few years, but even the darker, more uniform depths of soil have time scales of material turnover by chemical reactions of only hundreds of years. These turnover times are finger snaps relative to the cycles of chemicals in the rocks just below the soil, which take millions and even hundreds of millions of years. (These rocks should therefore be considered as below the biosphere.) The various "rapid" dynamics in the blendings of air, ocean, and soil make the biosphere a truly unified "thing."

Yet no matter how well stirred, the biosphere is not completely closed to entries and exits. Take carbon, for example. Dave the carbon atom was not always in the biosphere. We have seen him about to enter the atmosphere from inside a glass of beer upon his conversion from ethanol to CO_2 inside a human body. We have seen how, several months earlier, he entered a barley leaf in the form of CO_2 from the atmosphere, and then how he fairly quickly went into the grain as starch. So Dave goes in and out of various parts of the biosphere, as parts of cycles of carbon circulating all that time completely within it. But how did Dave enter the biosphere?

Dave's Origin

When planning this book, I decided that I would not have Dave the carbon atom come into the biosphere as part of a fossil fuel (say, out of an oil well) and then combusted in a car's engine. David Keeling had a passion for studying the biosphere. He discovered how plants shift atmospheric CO_2 levels, and he was concerned about humanity's unremitting dependence on burning ever more fossil fuels. Thus, I thought Dave the carbon atom's entry should be natural. (Soon I will introduce some carbon atoms that originated from a coal mine or an oil well.)

Dave came into the biosphere scene when a bit of limestone rock dissolved in rainwater 32,000 years ago.[1] The coldest part of the on going Ice Age had not yet been reached, but even then sea level was 80 meters lower, because across North America, Europe, and Asia much water was locked up in continental ice sheets that resembled those of Antarctica and Greenland today. Humans had already existed in their fully modern biological form as *Homo sapiens* for at least 100,000 years. But culturally, an explosion of creativity was just beginning to swell. It was destined to alter the biosphere forever. Some early harbingers of this explosion are evident in the caves of present-day southern France. In the Chauvet cave, magnificent images of rhinoceroses and lions

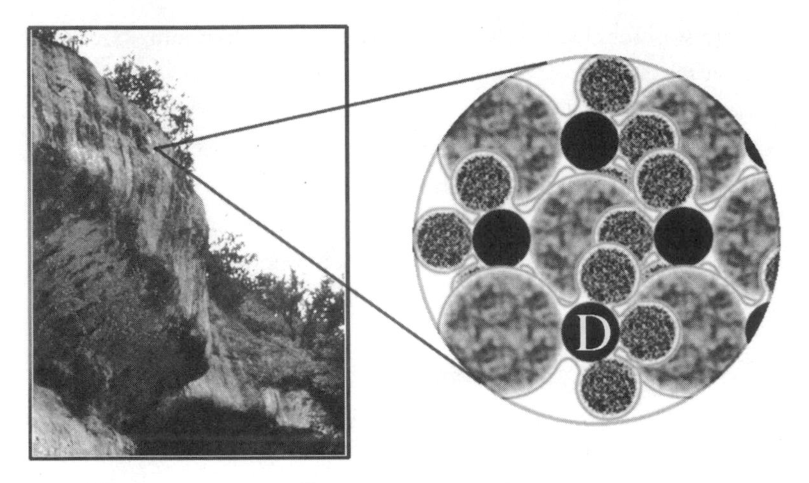

Figure 4.1 Carbon atom Dave sits at the surface, in a lattice of calcium carbonate inside a limestone cliff of southern France 32,000 years ago. He is about to be washed out by rain and into a river. Calcium carbonate is a repeating crystalline lattice consisting of a basic unit of one atom of carbon (black), one large atom of calcium (the big one), and three atoms of oxygen (mottled dark gray).

were painted on deep cave walls. In Germany, around the same time, someone made a sculpture of a man (perhaps a shaman) with the head of a lion.

Dave was then on the surface of a rock on the side of a limestone cliff in the Dordogne Valley of France, where many inhabited sites from the Upper Paleolithic are preserved. Dave's neighbors included an atom of calcium and three of oxygen, and those ratios held throughout the interlocked crystals of the rock Dave was part of: one atom of carbon, one of calcium, and three of oxygen. Dave's molecular home was calcium carbonate ($CaCO_3$), the common mineral known as limestone (figure 4.1).

Dave had been with the same neighboring atoms for 28 million years—ever since the geological period called the Oligocene, during which the earliest small horses and many types of grasses originated. But before this incomprehensively long crystalline entombment, during a "brief" 100,000-year period within the Oligocene, Dave had been free and active in the biosphere. His travels ended during that 100,000-year period when he got incorporated into the shell of a microscopic marine creature called a coccolithophore. When that coccolithophore died, it sank into the sediments at the bottom of the waters along a continental shelf and was buried by the continuous thin rain of detrital materials that drifts down to the ocean floor.

Sediments weighed down on Dave and his neighbors and eventually welded them into limestone. In time, the land shifted. With the general uplift of southern Europe, which built the Alps, Dave's crystalline tomb became elevated, putting him in a sedimentary layer of rock well above sea level. Rain eroded the minerals bit by bit, forming valleys. Millions of years went by. The crystal neighborhood that contained Dave came closer to the surface as previous outermost slips of mineral were weathered away by rain, which dissolved them and washed them into the active biosphere. Eventually, 32,000 years ago, Dave and his nearest neighbors were right at the surface of the limestone cliff. Some rainwater rinsed across and pulled Dave and the three oxygen atoms apart from the calcium atom. Dave was again loose in the biosphere after millions of years of crystalline imprisonment. Carried downward in the drops, and now in a carbonate ion, Dave eventually entered the Vézere River, then the Dordogne, and then the ocean. Once in the ocean, he would not necessarily stay there very long. He could become part of the air, or of soil, or of a plant, or of a woolly rhinoceros like the one that was being painted on a cave wall at the time of his release.

Fossil-Fuel Carbon Atoms Join the Biosphere

According to most anthropologists, Paleolithic humans probably spoke a complex language. They probably had sophisticated social structures. We know that they used a wide range of tools, including fine bone needles and hunting spears tipped with carefully crafted flint blades. They probably had myths, an awareness of past and future, and sufficient mental abilities to make plans essential to their daily lives throughout the cycles of the seasons. They had complex burial customs. It has been inferred that the deep foundations of innate human psychology had already evolved, including the ability to act on the assumption that others had minds with beliefs that could be true or false and the ability to assume that others held goals that were reasons for their behaviors. The statue of the lion-headed man seems to indicate an ability to combine concepts and construct a mental world that was not just an imitation of nature.[2] They were already using external sources of energy—for example, by making fires that unlocked the energy of the carbon atoms in wood. In the course of 32 turnovers of the ocean, each taking about 1,000 years, their descendants would eventually learn how to pull fossil fuels from deep under the ground on a huge scale and how to burn the fuels to create the fiery transformations of matter that would signal the full breaking of the cultural and industrial tidal wave upon the biosphere.

For all of the 32,000 years since Dave entered (or, more properly, reentered) the biosphere after the dissolution of an Oligocene limestone crystal, he had traveled far and wide, high and low, in and out of various creatures large and small, had floated in the darkest ocean depths, and had felt warm sunlight while inside a leaf. He wafted in the atmosphere for years at a time during the construction of some of the earliest agricultural villages of the Anatolian Plateau of present-day Turkey. For 300 years he was in a molecule of cellulose inside a tree in what is now Costa Rica. He was blown through the windmills of

Renaissance Holland. He spent 100 years in the soil of the pampas of Argentina while England experienced the Industrial Revolution. And in the early 1960s he passed through the infrared gas analyzer at Mauna Loa.

Let us now turn our attention to three other carbon atoms that entered the biosphere's chemical cycles at about the time that Dave went through the gas analyzer.

Enter Coalleen

A month earlier before she entered the biosphere's chemical cycles, carbon atom Coalleen was dug up as part of a chunk of black coal from an open-pit mine in the central region of Inner Mongolia. The chunk was loaded with tons of others on a smelly diesel truck and bounced along rough roads toward a coastal town in China, where the chunks were pulverized and then pressed into hand-sized bricks of coal bound for furnaces.

In early winter, Coalleen was used to heat a restaurant in northern China. Her brick was tossed into a small furnace with hand-held tongs. Ignited by flames from the remnants of the previous brick, her brick became, in turn, the actual source of those flames. The heat caused Coalleen and her atom neighbors to vibrate. Their electrons swung outward and then rebounded inward many times as oxygen was swept into the furnace through the intake. The oxygen became hellishly hot. It greedily grabbed some of the vibrating electrons in the molecules of coal, and with its electrons formed bridges with the carbon atoms that had supplied the vibrations. As a result, the carbon atoms entered a state of lower energy, having had their electrons partially taken over by the oxygen atoms. In the process, heat was liberated and carbon dioxide formed. This waste gas was propelled up the furnace's chimney by a rising plume of hot exhaust fumes. As a breeze dispersed the exiting fumes, Coalleen entered the atmosphere within a new molecule of CO_2. She had not been in the active biosphere for more than 200 million years.

To think about the formation of coal, think of tree trunks, tough bark, and thick roots. Think of swamps full of cypress-like trees with their roots arching in the air and supporting their trunks like the flying buttresses of French cathedrals. Above all, think of giant ferns and mosses.

Ancient trees, giant ferns, and gigantic mosses, like the bodies of all things living today, were full of carbon and dense with structural molecules of cellulose, hemicellulose, and lignin. The low levels of oxygen in the water-logged soils of swamps and bogs limited most of the usual opportunities for bacteria and other soil creatures to feast upon and thereby degrade the dead bodies of once-living things that fell into those damp, slimy ecosystems. The slowness of decay in peat bogs, for instance, provides the reason for many a whoop of joy from archeologists who discover a bog man or woman preserved after thousands of years, even down to the clothing, soaked, soiled, and in tatters, the body having been thrown into the bog after a ritual of human sacrifice.

Certain times in Earth's past have been better than others for the creation of the rock deposits we call coal. Though wet times seem to have been more propitious for the formation of coal, sometimes coal deposits do derive from geological layers laid down when the climate was drier, and thus the science of the origin of coal is still a work in progress. But it is well established that coal originated when dead terrestrial vegetation failed to be attacked by fungal and bacterial decomposers, which would usually have returned the dead plant's carbon to the atmosphere as CO_2. So the carbon in the molecules of coal was carbon removed from the biosphere that had not been fully degraded and recycled. This is one example of carbon burial, the technical term used by geologists for the downward exit of carbon atoms from the active biogeochemical cycles of Earth's surface biosphere.

This burial can last hundreds of millions of years. During all that time, the carbon is completely out of circulation and stays fairly much with the same atomic neighbors. There is some degree of chemical alteration in the layers of buried trees and other plant material that eventually become valuable coal seams. The buried organic material is pressed by the weight of new layers of mineral sediments above, as land subsides and erosion from elsewhere piles sand and silt into layer upon layer of sedimentary rock. The future coal seam is heated by the pressure of the new rock above and often by lateral forces as well, when rock layers are squeezed like an accordion and heaved into sinusoidal waves within mountains, as in the Appalachian Mountains of the eastern United States. These events drive some atoms from the future coal seam into gases such as methane. The coal also contains atoms of sulfur, varying with the specific ecological conditions of the original swamp or bog and the subsequent history of preservation of the organic material. With the exit of methane and other gases, the coal contains less hydrogen and therefore has a much higher fraction of carbon than the original plants had. The coal is compressed and homogenized into a fairly consistent solid dark brown or black. Depending on its age and its history of heating under pressure, the coal can pass, during what is known as coalification, through stages we somewhat arbitrarily designate as lignite, bituminous, and anthracite. But all these types of coal contain carbon, and from any of them we can obtain energy (figure 4.2).

In a sense, people want Coalleen and her atomic neighbors to be turned into CO_2 and released into the biosphere, because this frees energy that had been a part of the electron bonds of the atoms of carbon in the coal. This energy is then put to good use. In the United States,

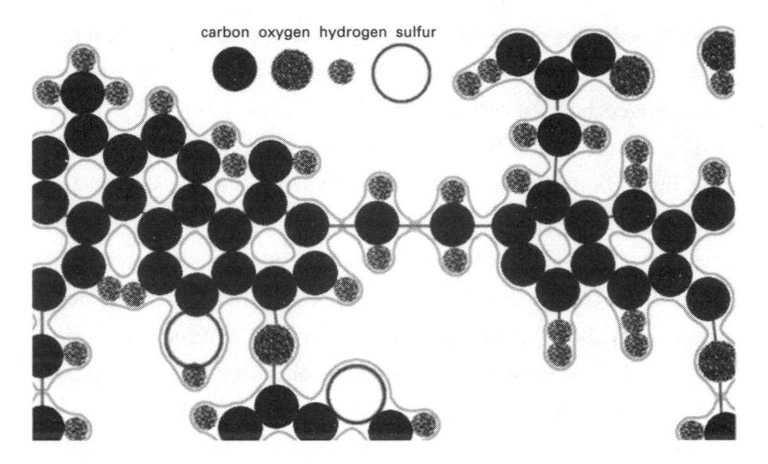

Figure 4.2 A generic proposal for a small portion of a molecule of coal. Coal is a mixture of many different kinds of molecules, so this is just one suggestion. Similarities to the cellulose molecule (figure 1.4) are obvious, and indeed cellulose was one of coal's main ingredients when it was alive as ancient trees and giant mosses. The combustion of coal in the presence of oxygen (in air) yields mostly CO_2, a little water, and some sulfur dioxide.

for example, about half of all electricity comes from burning coal. Coal was the first fossil-fuel energy of the Industrial Revolution, producing sooty skies and damaged lungs as well as power to turn wheels for manufacturing, usually by initially turning water into steam and then condensing the steam to drive a piston.

The energy released when coal is burned is no different from the energy obtained from burning wood, say in a campfire. And it is essentially the same reaction that your body undergoes when you create CO_2 as a metabolic waste by-product that you then exhale. In turning food into energy for the body, or in turning wood or coal into energy for external power, the basic reaction is the same: oxygen grabs electrons from an organic form of carbon, and the process generates CO_2 and energy.

Enter Oiliver

The second new carbon atom that we will track is named Oiliver. Like Coalleen, he entered the atmosphere at about the same time that Dave went through the gas analyzer at Mauna Loa. In a sense, we can say that Oiliver's debut in the biosphere took place the moment he was pumped up from way below ground in the giant oil field of Kirkuk in northern Iraq. Though not yet really a part of the biogeochemical cycles among air, water, soil, and life, at that moment Oiliver entered the financial cycles in which deadly serious money changes hands for unoxidized carbon atoms in liquid molecular forms.

Oiliver was pumped horizontally through a pipeline to a port on the southern coast of Turkey, then loaded onto a ship that took him to a refinery in the Netherlands, and then, in gasoline, burned in a French car on the highway between Paris and Bordeaux. Transformed and sent out the tailpipe, he entered the active biogeochemical biosphere as CO_2.

Oiliver's story, in general terms, is quite similar to Coalleen's. A biological form of carbon long buried and sequestered from any action within the biosphere is brought to the surface by humans and burned with oxygen. The energy difference in the two states of carbon, from oil molecule to CO_2, is used to push the pistons of the engine, which turn the crankshaft, which turns the wheels. It even powers (via transformation of mechanical energy into electricity by the alternator) the car's steering, its windows, its radio, and its seat warmers.

But there is one huge difference between Coalleen and Oiliver. The part of the biosphere in which Coalleen was buried was a plant, but Oiliver was buried in the feces of a zooplankton (a tiny swimming creature living at the surface of the ocean).

Algae and green bacteria (more properly called cyanobacteria) are phytoplankton, which grow by photosynthesis in the same way that land plants do. In seawater the phytoplankton are eaten by zooplankton—for instance, swimming copepods, tiny tadpole-shaped, crab-related things that occur in thousands of different species throughout the world's oceans. Included in zooplankton are also small larvae of mollusks and starfish, and among those are bacteria that feed on the wastes of all these creatures. There are, of course, bigger fish in food webs. Most large creatures generate gaseous, liquid, and solid wastes. As the solid wastes fall through the water column, they are consumed by bacteria.

Well, not consumed completely. Some part of this marine organic detritus, perhaps a hundredth of what was photosynthesized, makes it to the bottom, where worms and more bacteria in the sediments consume it. Well, not completely. Perhaps a tenth of that (now only a thousandth of the original amount) eludes the bacteria and is covered with more sediments, including silt and calcium carbonate shells from other organisms, and is permanently buried. Thus it happened that Oiliver was buried about 90 million years ago in a shallow sea near today's Middle East, in feces that had been released by a surface-dwelling copepod just a week before. He made it past the bottom-dwelling consumers and into a long, almost permanent burial in the sediments. Oiliver "went underground" for 90 million years.

Oiliver did have some subterranean adventures. The vast bulk of the organic detritus that is buried in marine sediments never becomes oil. Most becomes shale rock. But in some special areas, when the productivity of the plankton is enormously large, and the percent burial rate in the sediments is enormously large, the sediments contain chart-busting amounts of organic matter. Then, if the geological conditions are just right, the organic materials are buried under piles of overlying sediments and get heated at depths. This cooking drives off some of the hydrogen and even some carbon, and turns the material into a liquid. If the sediments are the right porosity, the liquid goop migrates until, sometimes, it finds a porous region and pools into an oil reservoir that is capped by impermeable rock above. The present industrialized world is pumping up black goop from such reservoirs of transformed dead plankton bodies at a rate of more than 80 million barrels daily, at 42 gallons per barrel, mostly to feed cars, trucks, buses, trains, and airplanes but also to make electricity, heat, and petrochemicals.

Enter Methaniel

When the interred land plants and marine plankton are pressure cooked by the geological underground chefs that turn these buried materials, over long time scales, respectively, into coal and oil, some carbon and hydrogen are driven off in the form of a gas, like steam from a boiling pot of soup. This gas product is primarily methane (CH_4), sometimes called natural gas (figure 4.3).

At Dave's moment of fame in the infrared gas analyzer, as Coalleen entered the atmosphere over China, and Oiliver entered above France, a mother of three in Canada was cooking spaghetti on a gas stove. Only a few weeks earlier, the gas, which had long been underground, had been vented along a vertical pipe up to Earth's surface, cleansed from any water vapor in the gas mixture, and, after other minor processing, shipped via more pipelines to temporary storage tanks, from which it was dispatched in a branching network of ever-smaller pipes.

Inside one of those molecules of methane in the flow of natural gas was a carbon atom we will call Methaniel. Like Coalleen and Oiliver, Methaniel had been absent from the biosphere for many millions of years. But we don't know-whether it was a terrestrial or marine creature that Methaniel was last in before his burial. That is because methane can come from either pathway. Sometimes we know a bit more. Methane

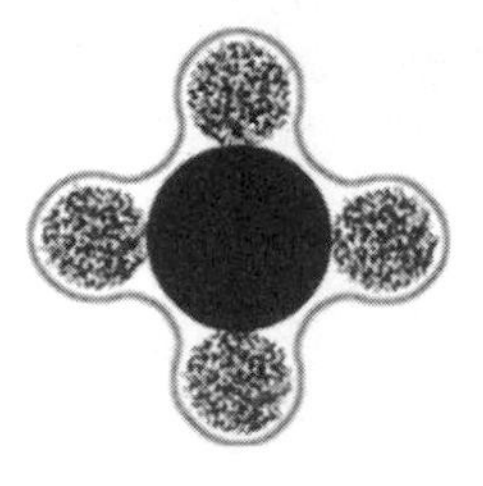

Figure 4.3 A molecule of methane, the primary constituent of natural gas. The central carbon atom is bonded to four hydrogen atoms—a simple chemical structure and formula, unlike the complex chemical hydrocarbon mixtures in coal and crude oil. The molecule is depicted here as flat, but in reality the hydrogen atoms form a three-dimensional tetrahedron around the carbon atom. Like coal and oil, natural gas generates CO_2 and water when burned.

near the tops of the inverted bowls of oil in the underground reservoirs came from the oil when the oil was cooked, and thus this methane was originally marine. Methane found in coal seams was distilled from the coal as it was processed. But sometimes, as in Methaniel's case, the gas has migrated so far underground from its original source that we cannot determine its source.

Natural gas contains carbon that can be burned in air, liberating energy as the bonds of carbon change from high to low energy. At the center of a tetrahedron of hydrogen neighbors, the carbon atom Methaniel had more energy in his electron bonds than he had in the post-combustion new bonds with the electron-greedy oxygen atoms in the liberated waste CO_2 gas. The difference in energy went into producing the stove's flame, some of which then went into boiling the water to cook the spaghetti for the Canadian family. So at that moment, in the early 1960s, at the same time as Coalleen's and Oiliver's incendiary coming out, Methaniel also joined the bustling activity of the biosphere.

Fossil-Fuel Inputs versus Natural Inputs to the Biosphere

All four of our named carbon atoms entered the biosphere after having been locked away and thus out of active circulation for many millions of years in their specific underground coffins of limestone (Dave), coal (Coalleen), oil (Oiliver), and natural gas (Methaniel). At first blush it seems that Dave is different from the others. For one thing, he came from rock rather than a fossil fuel. For another, his current spell in the biosphere began 32,000 years ago, whereas the three others have been with us for less than half a century.

Dave seems very different in another way, too. He is part of a cycle, not a one-way transformation. In our first look at a moment of Dave's life in the biosphere, he was in a molecule of alcohol in beer, soon to exit, after digestion, as CO_2 in the drinker's breath. Dave got into the alcohol as excreted waste from yeast that ate the maltose that had contained Dave in the beer malt, which in turn had been derived from starch in the barley grain, which had grown as part of the barley plant that incorporated Dave by pulling him in from the air when he was a wind-blown molecule of CO_2. In contrast, our first looks at how Coalleen, Oiliver, and Methaniel entered the atmosphere found them going from multi-million-year-old fossil carbon pulled up by humans from the deep Earth and combusted into CO_2 to serve various technological power systems and then exhausted into the atmosphere as industrial waste gas.

But this apparent difference between Dave and the others—that Dave is part of a cycle within the biosphere but the others are new atoms that just entered the biosphere—is illusory. After all, since the early 1960s, when the fossil-fuel-derived atoms of carbon entered the atmosphere, they have been circulating just like Dave, and it could have been one of them in the alcohol of contemporary beer. In other words, they too have been going through cycles within the biosphere.

To dig deeper into the story, and to perceive something that intuitively may seem fundamentally different between Dave and the three others, consider that I could have made Dave emerge from limestone in the early 1960s. Indeed, new natural carbon like Dave is being rinsed out of limestone, emitted by volcanoes, or outgassed from underwater vents and into the biosphere all the time. It is easy to retell the story so that the three fossil-fuel atoms would be older than the more recently released Dave. Suppose, for example, that Coalleen had come from coal in the earliest years of what the poet William Blake called industrial England's "dark Satanic Mills." Suppose Oiliver had come from the world's first commercial oil well, drilled in 1859 in western Pennsylvania by Edwin Drake. And Methaniel might have emerged from a natural gas well during World War II. Thus, all three could have pre-dated a more recent limestone birth of a different Dave.

So we have to move beyond any simple-minded analysis that says that natural carbon creates cycles but fossil-fuel carbon is from one-way, new injections. All atoms of carbon have their moments of origin into the circulation of the biosphere. Then, once in the biosphere, all atoms of carbon participate in circuitous cycles, in a cornucopia of looping pathways that I have barely touched on yet.

Carbon can enter the biosphere naturally from the dissolution of limestone rocks or from geologic vents, including volcanoes, and thus from underground. Or carbon can enter industrially from the combustion of fossil fuels, which themselves came from deep underground. These are only different ways for carbon to be released from its various kinds of ancient burial. Is there any crucial distinction, then, between natural carbon and fossil-fuel carbon?

It turns out that the difference between natural and industrial births lies in their birth rates.

The best way to estimate the total rate of natural release of carbon atoms into the atmosphere is to compute the global burial rate, which is better known. Averaged across sufficiently long time periods, the natural entry and exit rates are equal. We have measurements of the flow rates of calcium ions in rivers that enter the ocean, where many carbon atoms will eventually be buried, linked to atoms of calcium in calcium carbonate shells of microscopic organisms and reef-building coral. Marine biologists also include in this estimate the amount of organic carbon buried as part of the supply by rivers that carry, for example, the microscopic remnants of tree leaves, as well as the carbon that is buried as marine organic material, such as dead plankton bodies and other marine organic wastes. The carbon carried out as calcium carbonate shells dominates the numbers.

By estimating such numbers in the global carbon cycle, it is estimated that all natural sources produce a flux into the biosphere of about 400 million metric tons of new carbon per year.[3] How does this number compare to the injection of new carbon into the biosphere from all the various ways of combusting fossil fuels? The fossil-fuel input is known to a fairly high degree of accuracy. Once the amount of each type of fossil fuel used per year is known (say, how many barrels of oil, tons of coal, or cubic feet of natural gas), those numbers can be multiplied by the concentration of carbon in each, which makes it possible to compute the total amount of carbon exhausted as CO_2 for each fuel.[4]

As figure 4.4 shows, by 1900 humans werer releasing a flux equal to nature's input of carbon, and they did it almost exclusively with coal. By the early 1960s, the global flux of new carbon from fossil fuels was more than 2.5 billion tons of carbon per year. So at that time, the human-caused flux was about 6 times the entry rate of new carbon from nature's own processes.

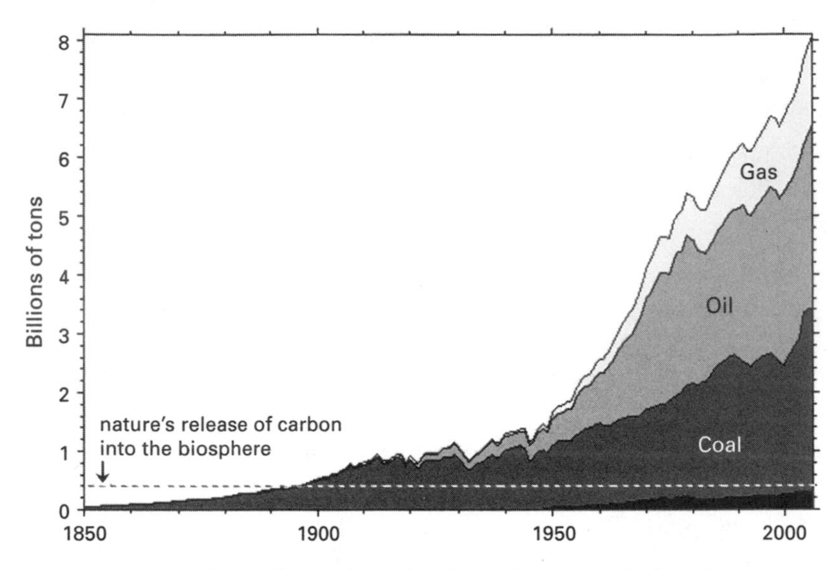

Figure 4.4 Billions of metric tons of carbon released into the biosphere as CO_2 from combustion of fossil fuels (natural gas, oil, and coal), per year, up to 2006. For comparison, the estimated value for the releases from natural processes, such as volcanism and the dissolution of rocks, is shown as 0.4 billion tons of carbon per year. The black region beneath the label "Coal" is from gas flaring and from the manufacture of cement, which takes limestone and heats it to drive off some CO_2.

There have been shifts in the relative magnitudes of the three fuels. In the early 1960s, coal was still dominant in terms of CO_2 emissions. Oil took over as number one sometime in the 1970s. Natural gas, almost insignificant in the 1960s, has grown into a substantial contributor.

With regard to the greenhouse effect, however, the atmosphere doesn't really care whether the CO_2 molecule contains a Coalleen, an Oiliver, or a Methaniel. The crucial number is the global total emission from fossil fuels. The numbers for 2006 indicate that the technological flux had reached 8 billion tons of carbon per year. Human activity outpaces nature's entry by a factor of 20. And the difference is growing.

I am reminded here of the words of Vladimir Vernadsky, a Russian biogeochemist of the early twentieth century. Vernadsky, one of the first to popularize the word "biosphere," recognized the power of life to affect Earth's chemical fluxes and said "Life is the most powerful geological force."[5] Humans, as an aspect of life, bring carbon up from deep below and put it into the biosphere at an ever increasing rate, which vastly exceeds the rate of global natural geological forces.

5

Carbon's Fluxes and Its Rate of Increase

The atmosphere is in constant exchange with plants, animals, and even the soil and the ocean. The CO_2 from each human breath spreads into the world's growing green leaves and down into the coldest ocean depths. Dave the carbon atom, now in a molecule of alcohol in beer, soon will reenter the atmosphere as CO_2 from one of those breaths. Carbon can exit the biosphere entirely, as Coalleen, Oiliver, and Methaniel did millions of years ago, when their previous incarnations in the biosphere were terminated when they were buried and became part of the underground stores of coal, oil, and natural gas. Therefore, some of the fossil-fuel CO_2 ejected into the atmosphere must now be somewhere else in the biosphere, or not even in it at all. Not all the fossil-fuel CO_2 remains in the atmosphere.

Are Fossil-Fuel Inputs Large Enough to Explain the Increase in CO_2?

How does the amount of fossil-fuel CO_2 that humans have expelled in fossil-fuel combustion compare to the increase of that greenhouse gas in the atmosphere? It is tempting to point to the fact that the industrial-derived input of CO_2 has been rising, then to the fact that atmospheric CO_2 has been rising, and to conclude that the former causes the latter. But we need numbers to be sure that this logical sequence is even a possibility to consider, even though the numbers by themselves might not be all we need for a definitive proof.

We should use a common currency for the comparison. Units of carbon are much better than units of carbon dioxide, for the following reason. The flux of carbon that beer-brewing yeast cells ingest during their consumption of maltose is not carbon as CO_2 but carbon in the form of the maltose molecule. The flux of carbon from a tree to the soil, when its leaves and other parts die and fall to the ground, is not as CO_2 but as carbon atoms in organic molecules. Thus, whenever we want to get quantitative about the fluxes in the global carbon cycle, units that are in terms of the masses of carbon in those fluxes allow us to look into the workings of the carbon cycle anywhere and compare any of its fluxes that contain carbon in multiple molecular guises.

A convenient unit for carbon fluxes on the scale of the biosphere is a billion metric tons of carbon, which was already used in figure 4.4. The total carbon emissions (as CO_2) from the triumvirate of fossil fuels is currently about 8 billion tons of carbon per year (8.8 billion American tons). The world's population in mid 2007 was 6.7 billion, so the per capita CO_2 flux is roughly 1.2 metric tons of carbon per person per year. (All tons are metric from now on, consistent with international standards for reporting carbon data.)

Let's add up the total contributions from all the world's countries since 1966, a few years after Dave's passage through the gas analyzer at Mauna Loa and after the entry of Coalleen, Oiliver, and Methaniel into the biosphere. The total fossil-fuel CO_2 emissions were then slightly more than 2 billion tons of carbon per year. During the 40 years up to and including 2005, the "contributions" to the flux of CO_2 to the atmosphere from the combustion of the three types of fossil fuel were as follows:

coal	86 billion tons of carbon
oil	98 billion tons of carbon
natural gas	36 billion tons of carbon.

Thus, the total carbon emitted into the air as CO_2 from all the fossil fuels used in technological processes, including a small but not trivial amount from cement manufacturing (about 5 billion tons) was about 225 billion tons.

How much did the concentration of CO_2 in the atmosphere increase during those 40 years? The atmosphere's CO_2 is measured in parts per million. Using calculations for the mass and composition of the atmosphere, we can convert that to billions of tons. It turns out that each part per million is equivalent to a global atmospheric mass of 2.13 billion tons of carbon. Therefore, if today's CO_2 concentration is about 385 ppm, our atmosphere now has about 820 billion tons of carbon circulating as CO_2.

Now we can figure the mass of carbon in the observed increase of CO_2. Over the 40 years, the CO_2 in the air increased by 61 ppm. Using the conversion factor above, that's an increase of about 130 billion tons of carbon as CO_2. So humans put 225 billion tons of carbon into the atmosphere, and the amount of carbon in the atmosphere increased by 130 billion tons.

The mere fact of the mismatch in these numbers does not allow us to dismiss fossil-fuel carbon. Clearly more is going on than the dumping of all this carbon into the atmospheric bowl. But it is evident that fossil-fuel carbon could be the source of the increase in the atmosphere's CO_2. Note that the flux ejected up into the air is greater than the observed increase in the air. We would be unable to pin the guilt on fossil fuels only if the increase in the atmosphere were the larger of the two numbers. In principle, clearly, the fossil-fuel release can account for the increase if some of the release each year ends up somewhere other than

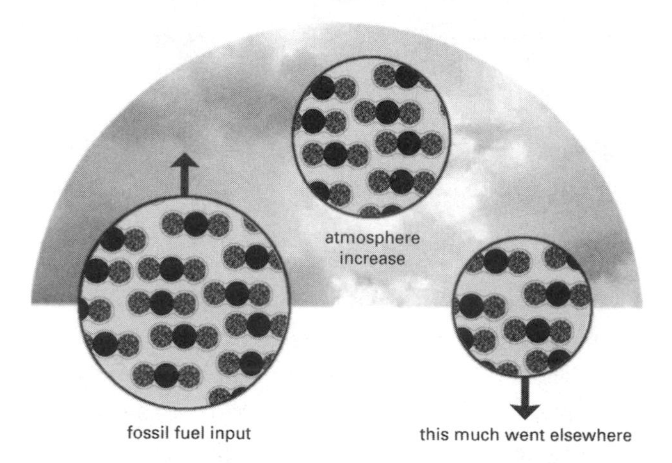

Figure 5.1 The carbon balance for the atmosphere from 1966 to 2005, with the amounts relative to the areas of the circles. Fact 1: By burning fossil fuels (plus a small contribution from cement production), humans shot 225 billion tons of carbon as CO_2 up into the atmosphere (left circle). Fact 2: The atmosphere's CO_2 content grew by 130 billion tons of carbon, as observed at sites such as Mauna Loa and the South Pole (middle circle). Those numbers imply that during this time period a net amount of 95 billion tons of carbon as CO_2 exited the atmosphere (right circle).

in the air, even though it all starts out in the air. In the 40 years we are considering, the increase in atmospheric CO_2 was 58 percent of the total amount that all human technological processes had put into the air. The rest—42 percent, or about 95 billion tons of carbon—must be somewhere else in the biosphere. (See figure 5.1.)

If a much higher fraction of CO_2 had exited the air, we could put off concerns about the amplified greenhouse effect. And if less had exited, we would already be more worried. As it is, the news is both good and bad: only about half of the CO_2 we inject into the air stays there, but enough stays there to drive a rate of increase that quickly accumulates into numbers that lead many to believe that serious changes are occurring.

Dave and the Fossil-Fuel Carbon Atoms All Leave the Atmosphere

After Dave went through the infrared gas analyzer in 1963, he traveled the globe on complex circuits of air currents. He passed back and forth between the northern hemisphere and the southern hemisphere a few times. Then, in 1967, he went down into the ocean by shuttling right across the surface of the water. The motion of tiny currents tunneled him down between the water molecules, and very quickly his CO_2 molecule joined with a hydroxyl ion (born from the occasional, random separation of the water molecule into its own ions) to become the dissolved bicarbonate ion HCO_3^-), the most abundant form of carbon in the ocean. (See figure 1.3.)

Coalleen, Oiliver, and Methaniel all started their travels in the atmosphere in 1963 and so were in the air with Dave. They, too, rode around in the gaseous giant horizontal flywheels of high- and low-pressure systems, and they traveled upon easterly winds in the tropics and upon westerlies in the mid latitudes. But after only a year of airborne freedom, Coalleen passed through a stomate on the underside of a leaf of a strangler fig tree along the east coast of Australia and into the tree.

Oiliver stayed aloft until 1966, when he plunged across the air-sea interface, preceding Dave by a year to became one of the many bicarbonate ions in the sea.

Methaniel had the thrill of additional global circuits in his stay in the atmosphere, for he rode the air currents all the way until 1969, when he went into the leaf of a tundra plant in northern Siberia.

Were these spells in the atmosphere for our atoms anomalously brief? If they were typical, then it would seem that the fossil-fuel CO_2 leaves rather quickly—perhaps too quickly, in view of the numbers just cited. Is there any big-picture evidence we can look at to see if the fluxes between air and, say, plants can be so large that most fossil-fuel carbon atoms are shunted out of the atmosphere in only a few years?

A Clue from the Annual Ups and Downs

I noted earlier that the annual oscillations of CO_2 in the northern hemisphere are due to vegetation, and that David Keeling had made this discovery. For example, in figure 3.2 the CO_2 measured at Mauna Loa goes down by about 7 parts per million from May until September or October, and then up by about 9 ppm from the autumn low until the May of the following year, creating a net rise of about 2 ppm per year in recent years. The up and the down parts of the annual oscillation, in absolute magnitudes, are each several times the average annual rise.

In figure 3.4 we saw that the seasonal oscillation for Barrow, Alaska, is greater still, about double that of Mauna Loa. Furthermore, the oscillation was either small or negligible at the sites in the southern hemisphere—for example, at the South Pole (figure 3.3) or American Samoa (figure 3.4).

Certain places seem to have a suction pump that operates during summer and pulls CO_2 from the atmosphere. During winter, the pump works in the opposite way, expelling CO_2 up into the atmosphere. Can we ascertain by logic what these pumps are, without taking Keeling's or anyone else's word for it?

There are three possibilities. We can logically eliminate two of them.

The first possibility is human industrial activity. Most of the industrialized countries that burn the major portion of the fossil fuels are in the northern hemisphere. It might be the case that wintertime activities in the northern hemisphere create more CO_2, because homes, offices, stores, and factories all require heating.

The monthly data for fuel use show more burning of natural gas and fuel oil in winter. But in summer, pleasure driving and the increased output of power plants to supply air conditioners with electricity nearly make up for the absence of the need for heating fuels. Still, overall, there is a slight seasonal cycle in the use of fossil fuels. But even a large seasonal cycle in the use of fossil fuels (if there were such a cycle) could not produce the ups and downs of the Mauna Loa or the Barrow data. Any seasonal cycle in fossil-fuel use would only create a cycle in the ongoing rate of rise in the atmosphere—for example, a smaller increase in summer than in winter. But the CO_2, in this hypothetical scenario, would always be climbing, all year round; it would never exhibit the net decrease that appears from May to September or October. We need to search elsewhere for an explanation of the seasonal cycle.

The second possibility implicates the oceans. Perhaps there is more of a chance that, when CO_2 molecules such as those that contained Oiliver and Dave pass from air to ocean, as they did when they respectively left the atmosphere in 1966 and 1967, they do so, for whatever reason, at greater rates in summer than in winter. And perhaps more CO_2 has a tendency to come out of the ocean and pass up into the air during winter.

Temperature alone cannot help us, because cold water holds more gas than warm water does. (Think about the fizz from warm soda or warm champagne.) But perhaps there is another factor, such as photosynthesis by tiny marine phytoplankton. Globally, marine photosynthesis is known to be nearly equal to that of the land's vegetation. But here is the crucial expectation that can be tested if the ocean is involved. No matter what cause we presume to create a back-and-forth net seasonal flux between air and seawater, we should see much larger oscillations of CO_2 in the southern hemisphere. That's because the southern hemisphere has a substantially larger fraction of ocean surface than does the northern hemisphere. In the northern hemisphere, the ocean's area is about 50 percent more than the area of the continents. In contrast, in the southern hemisphere, the area of the ocean is four times the area of the land.

There is no other significant difference in the seas of the two hemispheres. They have about the same amount of salt concentrations. They both have nutrients, plankton, fish of various sizes, and even whales. They both experience seasons in which the sun rises and falls in intensity, and more so with increasing latitude. The oceans of the northern and southern hemispheres differ only in size.

As noted, the data are striking for the lack of seasonal oscillation in the CO_2 of the southern hemisphere. So the oscillation in the atmosphere of the northern hemisphere cannot be the result from a seasonal shift in the passage of a net flux CO_2 from air to water in summer and then from water to air in winter.

By this line of reasoning, in which we have eliminated two possibilities, we can now conclude, as did Keeling (who also used isotope evidence), that the cause of the seasonal cycle in the northern hemisphere air is the only conceivable third candidate: the seasonal cycle of terrestrial land life.

Plants, as they grow in summer, pull CO_2 out of the air. Northern-hemisphere plants, collectively, are the suction pump.

In winter, photosynthesis is very low in the mid to high latitudes (continuing only in the evergreen trees, and even in them at a much lower rate than in summer). But we and many animals keep exhaling CO_2 all winter long. And so do the bacteria in the soil. They are slowed down by the cold, but they stay alive and, therefore, by metabolic necessity keep churning out CO_2 as a waste gas from their respiration. Their excreted CO_2 percolates out of the soil and passes into the atmosphere.

Let me elaborate a bit on a fact mentioned above. Right away we can make one definite statement about the seasonal oscillation and the power of trees, grasses, and herbs: When the oscillation is in its annual downward swing, the flow rate of CO_2 into plants, in sheer magnitude, is greater than the steady injection of fossil-fuel CO_2 that creates the upward trend.

It is easy to compare the slope downward during summers with the average slope upward of the overall trend. Figure 5.2 shows that the summertime drop is very steep. The rate downward is about 6 times the rate of the average trend upward. Thus, during peak summer months, the biologically driven northern-hemisphere removal mechanism of CO_2 is 6 times the fossil-fuel rise. Thus, natural fluxes, at least during certain months, can dwarf the fossil-fuel release.

The seasonal cycle is strongest at high latitudes. It is stronger, for example, at Barrow, Alaska, than at Mauna Loa. But because the oscillation is close to zero at the equator, it turns out that Mauna Loa is a good northern-hemisphere average, so we can use Mauna Loa and not require any additional fancy averaging across the monitoring sites of the entire northern hemisphere.

For those who want to cheer the power of nature, the situation is even better than the factor of 6 during summer. The seasonal oscillation derives not just from plants during the summer and bacteria, say, during winter. The oscillation results from the changing seasonal differences in the net flux of what is usually a bi-directional exchange between ecosystems and the atmosphere.

Soil bacteria live plenty well during the summer, too. In fact, they churn forth CO_2 across their membranes with greater ferocity in

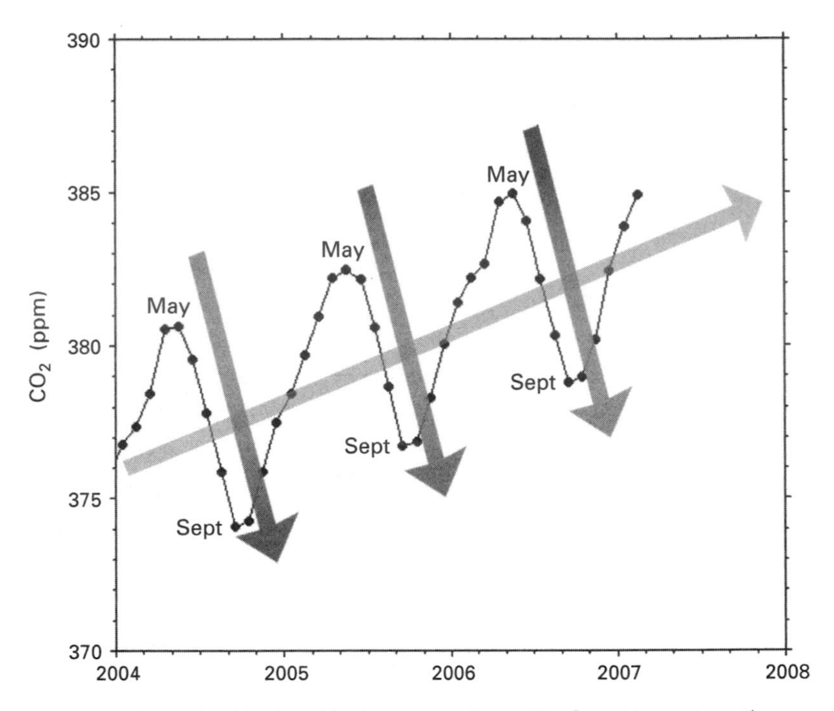

Figure 5.2 Monthly data (dots) on atmospheric CO_2 from Mauna Loa. The upward trend of the rise from fossil-fuel CO_2 is apparent. But between May and September each year the CO_2 decreases, which is representative of an overall decrease in the northern hemisphere atmospheric average. This dramatic decrease comes about because summertime vegetation grabs and incorporates carbon from the atmospheric CO_2 pool at a rate that more than counters the steady emissions of fossil-fuel CO_2.

summer than in winter, because their little bodies are then warmer and therefore more metabolically active. Thus, the drop we see in CO_2 during the northern hemisphere's summer is the difference between the carbon that plants take up in photosynthesis minus the output in respiration from all fungi, worms, insects, birds, mammals, and bacteria (by far the largest contributors, which is why I usually focus on them).

Averaged over the course of the year for both hemispheres, and including the tropics where photosynthesis occurs year-round, the total annual uptake of carbon by global terrestrial photosynthesis is on the order of 60 billion tons of carbon per year. This dwarfs the human fossil-fuel emissions (about 8 billion tons per year). So does the presumed 60 billion tons of carbon per year released from global terrestrial life during respiration.

But this analysis still leaves us with a problem in our quest to figure out where 40 percent of the fossil-fuel CO_2, annually, on average, goes. Back and forth fluxes between atmosphere and living things from seasonal and nonseasonal ecosystems don't help us get rid of some of the fossil-fuel CO_2, if those back-and-forth fluxes are balanced in their annual averages. Yes, the back-and-forth fluxes exchange the fossil-fuel CO_2 carbon atoms that were put into the atmosphere for other atoms now in the atmosphere that came from ecosystems (say, as dead trees decayed). Fossil-fuel carbon atoms are taken out of the air and other carbon atoms are put in. But mere replacement does not create a net removal of CO_2 from the atmosphere. For a candidate for a global CO_2-removal mechanism that is obviously better, we must turn seaward.

The World Ocean

If we could analyze the world ocean at the required level of detail and track the carbon atoms in the CO_2 that you or I exhale from each controlled collapse of our lungs, within a few years every spot on the ocean's surface would contain molecules, mostly bicarbonate ions, that themselves contained carbon atoms that had been in the CO_2 molecules that we exhaled. And that is from every one of our exhalations, essentially across our entire lives. Within 1,000 years, an oceanographer could send could a jug down into the deepest depths (say, 3 or 4 kilometers beneath the surface)—anywhere—and pull up water with a bicarbonate ion whose carbon atom came from any single exhalation that you or I made over our entire lives.

The carbonated bubbles in the beer contain CO_2 molecules that were excreted by the yeast in that other waste pathway that Dave, in the alcohol molecule, just happened to miss. Those bubbles rise in the beer and pop across the surface and pass up into the air because the concentration of CO_2 in the beer is greater than in the air, which is why we experience the beverage as carbonated. But the process could work the other way around. Are we "carbonating" the atmosphere with our fossil-fuel releases, and does that drive some of the CO_2 down into the ocean? In 1966, as was noted, Oiliver traversed the border between air and water and went into the ocean. Dave took the same plunge a year later.

But after Dave and Oiliver entered the ocean, their paths were quite different. In fact, there are probably more fundamentally different paths for a carbon atom to follow in the ocean than on land. In 1970, after 4 years in a bicarbonate molecule dissolved in seawater in the South Pacific, Oiliver went back up into the atmosphere as CO_2, simply reversing the gas exchange process to return him to the atmosphere. Dave also became part of a bicarbonate ion, but then, in 1971, was taken up into the tiny body of a marine plankton called a diatom, one of the many species of photosynthesizing plankton. This one grew during the spring bloom in the North Atlantic. So Oiliver went right back into the air, while Dave was to stay awhile in the ocean "pool" for longer.

Figure 5.3 CO_2 and other gases move in and out of the ocean across its surface. Here a technician on a oceanographic research voyage prepares a special bottle to be used to sample water near the surface. The water will be analyzed to provide data on the rate of gas exchange.

What is the total flux from air to ocean, or from ocean to air?

Oceanographers call it gas exchange, and indeed it occurs in both directions (figure 5.3). There is no simple way to visually demonstrate its power, as we could do with the terrestrial exchanges that cause the seasonal CO_2 oscillations in the northern hemisphere. But the magnitude of gas exchange, because of its importance, has been the focus of many experiments.

Radon, a radioactive gas that occurs in the ocean as a result of the breakdown of dissolved natural uranium atoms, is depleted at the ocean's surface because it flows up into the atmosphere. The magnitude of the depletion can be measured and the gas exchange rate calculated for radon, and then recalculated for CO_2, using formulas that take into account the different molecular weights of the gases. In other measurements, sulfur hexafluoride, which does not occur naturally, has been added to the surfaces of lakes and ocean regions, and its dispersal into the air monitored to give a value for the rate of gas exchange (which again must be recalculated using well-honed principles of the physics of gases). And there are other techniques.

Overall, the experiments and measurements show the following. The flux of CO_2 between atmosphere and ocean is huge: about 100 billion tons of carbon per year. That's in both directions, approximately, and such numbers are 12–13 times the flux of 8 billion tons per year in the entry of fossil-fuel CO_2 into the atmosphere.

Of course, what we want to determine, ultimately, is whether there is a net flux of CO_2 from the air to the ocean that removes some of the fossil-fuel CO_2 we put into the air. In this way, we might expect that the atmosphere and ocean together act like a reverse bottle of beer. We are carbonating the atmosphere, and some of the fizz is moving down into the ocean. But before we get to that, let us consider some further questions raised by the presence of such large fluxes in the natural systems of the biosphere.

A Question about the Large Fluxes of Nature

About 100 billion tons of carbon as carbon dioxide pass both into and out from the ocean per year. On land, 60 billion tons of carbon are incorporated into the bodies of photosynthesizers and are emitted from respirers, such as vertebrates, insects, and most bacteria. But plants respire, too. And so we must include the flux of carbon that plants respire (and therefore will have taken in) to bioprocess the sugar formed by photosynthesis, which was necessary to re-package Dave into starch in the barley seed. This gross photosynthesis is twice the net photosynthesis, so the amount of carbon that actually passes into plants is more like 120 billion tons per year. (See figure 5.4.)

Such large fluxes confront us with a question. To what extent can we assume that the fluxes of nature form a predictable background to the CO_2 problem? Specifically, if any of the two-way exchanges of nature get out of balance—say, either the two-way biological fluxes on land or the two-way gas exchange fluxes across the ocean's surface—then the exchanges could become net fluxes in one direction or the other. They could, at least from the analysis so far, become net fluxes to the atmosphere. Or net removal mechanisms.

In fact, today one or more of them must be a net removal mechanism with respect to the atmosphere, because, as figure 5.1 shows, not all the CO_2 that we put there stays there. But could the exchanges in nature get out of balance in a big way, and all by themselves?

The implications of such a possibility become clearer by evaluating how much carbon is in the various places—the "pools" or reservoirs—of the global carbon cycle. Inventories have been taken. Although the numbers are subject to some uncertainty, the general values are well enough established for our purposes. Rounded off, they are as follows:

atmosphere	800 billion tons of carbon in CO_2
land plants	600 billion tons of carbon in biological tissue
ocean life	2 billion tons of carbon in biological tissue
surface ocean	800 billion tons of carbon, mostly in bicarbonate ions
deep ocean	35,000 billion tons of carbon, mostly in bicarbonate ions.
soil	2,000 billion tons of carbon in biologically derived organic detritus.

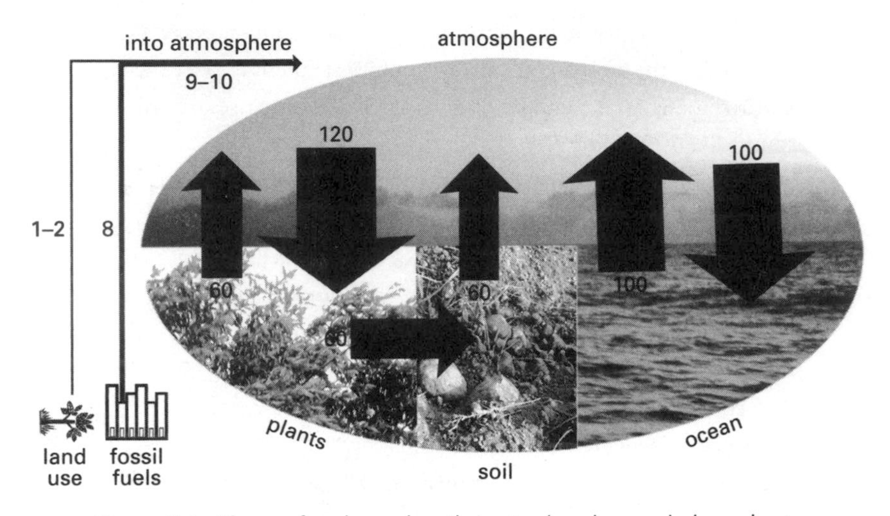

Figure 5.4 Fluxes of carbon when the natural cycles are balanced: air-ocean gas exchange of CO_2 (100 billion tons of carbon each way per year); photosynthesis by terrestrial plants that turns CO_2 into organic molecules of life (120 billion tons of carbon per year); respiration by plants to build more complex molecules inside their bodies and release of CO_2 (60 billion tons of carbon per year); transfer of carbon (primarily by plants but also by animals) as detritus into the soil (60 billion tons of carbon per year); and respiration by soil organisms (mostly bacteria), which releases CO_2 into the soil and then up into the air (60 billion tons of carbon per year—for simplicity this includes respiration by land animals, including insects and mammals, of about 5 billion tons of carbon per year). For comparison, the flux of CO_2 from the combustion of fossil fuels is shown (8 billion tons of carbon per year in 2006), which goes into the atmosphere but which is here kept outside the system, because when it is inside (as it really is) the other fluxes are no longer in perfect balance. Also shown is an estimated flux from land-use change. Not shown are fluxes of carbon inside the ocean—for instance, photosynthesis by phytoplankton (40 billion tons of carbon per year) and respiration by marine organisms.

What would happen if the flux between land ecosystems and atmosphere were to become unbalanced by approximately the same amount as the fossil-fuel CO_2 input, about 8 billion tons a year? Suppose this ecosystem imbalance would go into the atmosphere. There is enough carbon in land plants and in the soil (2,600 billion tons of carbon, total) to supply such an imbalance for more than 300 years. The ocean pool has even more carbon that could feed an imbalance. The net flux in air-sea gas exchange could be several times the fossil-fuel flux and there still would be enough carbon in the ocean to supply that imbalance for 1,000 years.

Now, other factors are going to come into play on these time scales, factors that in the first place created the respective amounts in the pools (or bowls) for given degrees of fluxes. But we should not assume without proof that nature is in balance and humans are not. Nature is not necessarily in balance. We should not hastily claim that nature follows some biosphere scripture for good behavior. Some of these fluxes must be imbalanced, because the amount of CO_2 added to the air from fossil-fuel combustion surpasses the measured increase in the air over 40 years. But how do we know that one part of nature is not contributing to the increase, say by putting an extra flux into the atmosphere and that another part of nature is taking out portions of both the fossil-fuel flux and also a possible extra flux from one imbalanced part of nature?

In fact, scientists who develop budgets for the carbon cycle know that human activities have created a net flux to the atmosphere in addition to the flux from fossil fuels. This second flux is from "land-use change" (figure 5.4)—an anthropocentric expression when applied to "uses" such as the conversion of pristine tropical rain forests.

As is clear from the numbers above, because land plants and soils are two substantial pools of carbon, fractional changes of the total land ecosystem pool—of its plants or soils—can be significant.

One analysis of changes in the carbon fluxes associated with changes in land use goes back all the way to 1850.[1] Then great forests were being cut, especially in the temperate regions of the northern hemisphere as human populations expanded. The clearing of forests emits CO_2. If forests are cleared by burning them, they release CO_2 gas immediately and in large amounts. If the forest is harvested, the small branches and twigs that are left behind to rot, as well as the dead roots from the cut trees, decompose rather quickly. Some of the forestry products go into wood, which doesn't put the CO_2 immediately into the air. Eventually, however, as the wood becomes trash, bacteria decompose it and push CO_2 first into the soil and then into the air.

The budget for all these changes can get quite complicated, because carbon in wood that was cut in 1750 for a piece of French furniture can still be active in use as furniture and not in the air. And sometimes, after forests are cut, they begin to regrow. So the timings of the releases of CO_2 during deforestation, as well as the increases in the local carbon inventory during forest regrowth, must go into the accounting for the total global carbon inventory during land-use change. And one more piece of this budget—and not a trivial one—is the change to soil carbon when the soil is put under the plow. Cultivation systematically lowers the carbon content of the soil by warming and aerating it. Both processes increase the activity of the soil's bacteria, those little but globally powerful subterranean hydrants of CO_2.

In land-use changes, I emphasize, one must also account for regions in which agricultural land is reverting to forests, as is happening in the northeastern United States. In these places, carbon is entering storage in the cellulose and lignin of the growing trees, which take CO_2 from the air to create their bodies.

It is indeed a complicated matter to add up the pluses and minuses in all these sites of land-use change, each with their own uncertainties, all over the world. One estimate pegs the present-day flux to the atmosphere from land-use changes at about 1.5 billion tons of carbon in the form of CO_2.[2] There is some controversy about this because of the complications, and the number could be lower or higher. (For this reason, I have shown a reasonable range in figure 5.4.) Nevertheless, shifts in land use demonstrate the fact that changes in the pools of carbon that connect with the atmosphere do occur. Could those pools also change as a result of natural processes in the biosphere? One way to answer that is to get information about the global carbon cycle before humanity began to impinge on it.

6

Time Capsules in Ice

When Dave the carbon atom commenced his current stint of circuiting within the biosphere, 32,000 years ago, many other carbon atoms that had been Dave's close neighbors in rock did the same thing at virtually the same instant. For a reason soon to be revealed, we will call one of these new atoms Icille.

Dave and Icille had been in adjacent molecules of calcium carbonate, together in crystalline imprisonment for millions of years. When rain in Paleolithic Southern France washed Dave in a molecule out of the cliff and into watery solution, it also freed Icille. For a while the two continued traveling in close proximity, though now separately, each as a solitary carbon heart of a bicarbonate ion, with one hydrogen atom and three oxygen atoms as bonded partners (HCO_3 ; see figure 1.3). But soon random eddies of the Vézere swirled them ever further apart as it flowed into the Dordogne. A month later, when the Dordogne reached the Atlantic Ocean, Dave and Icille were many kilometers apart. And soon the ocean's waves and currents drove them apart by hundreds of kilometers. And after that, their molecular homes would diverge, Dave going down and Icille going up.

Dave's descent from the ocean's surface took place during the next few months. First his bicarbonate molecule gained a proton from the random jostling of various kinds of ions within the seawater, and Dave was hydrated into a molecule of dissolved CO_2. This chemical species was available for uptake by a planktonic cell full of green chlorophyll, actively growing by taking in CO_2 from the water to build the many carbon-based molecules of its tiny body. This green floater was next snatched up by a tiny free-swimming larva of a starfish, which in turn was gulped down by a herring, which in turn was snapped down the gullet of a spinner dolphin. A day later after that, the dolphin excreted Dave into the water in a particle that became attached to a clump of dead algae, which began wafting downward.

Bit by bit this so-called marine snow of detritus sloughed off into the water as it fell, and the organic material made available was feasted upon by the ever-present, countless marine bacteria that thrive on food delivered to the dark depths from the lit surface far above. About month later, 3 kilometers down, a bacterium attacked Dave's molecule itself within what remained of the falling marine snow. The microscopic bacterium took Dave's molecule (a fecal lipid), combined it with dissolved oxygen gas from the seawater to derive energy for its metabolism, then expelled Dave out its membrane and into the seawater as CO_2 waste, which soon reacted with the other chemicals in the salty seawater to put Dave, once again, into a dissolved ion of bicarbonate. But now Dave was 3 kilometers down.

Dave's Travels in the Biosphere during the Next 32,000 Years

Stuck in the deep, dark depths, where the ocean's currents are weak, Dave remained way below the surface within the southward-flowing giant tongue of seawater called the North Atlantic Deep Water for nearly 200 years. Inexorably and ever so gradually, he was lifted upward by the great currents of seawater that turn all the world's oceans over about every 1,000 years via intertwined conveyor-like horizontal crawls with feeder flows of descents and ascents. When Dave reentered the lit surface layer, he was in the southern hemisphere. Most of his subsequent travels over the next 32,000 years have been lost in unrecorded time. Nevertheless, our knowledge of the metabolism of the global carbon cycle allows us to say something that is at least statistically accurate about his whereabouts during these wanderings, against which those of Odysseus pale.

During about 200 individual intervals of time, Dave swirled in the atmosphere over land and sea. In the air, his chemical homes were almost always CO_2 molecules. But once (during the year 395 C.E., when the Roman Empire officially split into West and East) he was in a molecule of methane that had been generated by a soil microbe. These 200 aerial stays were almost always brief, on average about 4 years each. They began when Dave's carbon dioxide gas home popped out of the ocean, or was wafted skyward from the soil, or was ejected as a waste product of respiration from a terrestrial plant or animal.

During the 32,000 years, Dave inhabited organic molecules within green land plants about 100 times, almost always in a different species. Sometimes these vegetative stays were very short, only hours or days, because plants, as was noted in chapter 2, create waste CO_2 from about half their amount of carbon recently incorporated into sugar molecules from CO_2 during photosynthesis. This reoxidization allows plants to power additional reactions that put the other half of the new sugar carbon into more complex and permanent molecules. Sometimes Dave's stays in these more complex molecules within plants were also rather short-lived. For example, many times he was in flower petals that died and fell off, or in tender leaves eaten by caterpillars. But a significant number of Dave's stays in the plant kingdom were long. He lived within the woody tissues of tree branches, roots, and trunks 25 times, for a total of nearly 2,000 years. Once, 12,000 years ago, he had been in the bark of a towering sequoia tree in the Pacific Northwest for a straight spell of 600 years when he was swept away by the flood that followed the breaking of the ice dam that held back a giant lake in what is now western Montana.

Dave also spent time in various places around the world in soils of various colors and porosities, entering soil about 100 times as leaves, flower petals, or woody parts of trunks, as dung from caterpillars and other creatures, or as roots that died from drought or old age. Sometimes his stays in soil pools were short because he had barely arrived at the surface layer of litter when a bacterium would digest him, derive energy from him, and eject him as waste CO_2. Other spells in the dirt lasted years, even decades, and across a dozen intervals he stayed underground as organic carbon for centuries.

Within CO_2, Dave jumped from the atmosphere into the ocean about 100 times. Like his residences in plants and soils, some of his seawater stays were brief and others very long. Seventy times he was plunged below the ocean's well-mixed surface layer into what is called the marine intermediate waters. During about ten of these, his descent was in falling biological detritus, which could be organic (e.g., in the marine snow right at the start) or inorganic (e.g., in a tiny calcium carbonate shell made by a creature such as a coccolithophore or a foraminiferan). During the other 60 times, his home was a dissolved bicarbonate or carbonate ion and his downward travel was provided by the mixing of the ocean.

In about half of those times in which Dave took up residence in intermediate seawaters, he went even deeper, into layers such as the North Atlantic Deep Water. This passage came within falling biological detritus in only a few instances, because so much of the detritus that is formed in the surface food webs is biologically recycled in the surface and intermediate zones. For most of his descents into these truly deep waters, Dave's vehicles were the conveyor-like currents of seawater that cause cold waters in the high latitudes to sink downward. By whatever means he arrived, once he was in the deep waters his stays were lengthy, almost a millennium on average. All told, nearly three-fourths of his total time over the 32,000 years that we are surveying was spent in oceanic deep waters.

During all these spells of residence in the various pools, with their moments of passage and transformation between one pool and the next, Dave remained within the circuits of the overall biosphere. Eventually, in the early 1960s, while in air, he was shunted through the infrared gas analyzer near the top of Mauna Loa and played a part in the earliest measurements of the air's CO_2 content. At about the same time, the biosphere received Coalleen, Oiliver, and Methaniel as new atoms from fossil-fuel combustion.

Icille Gets Trapped as an Air Bubble in Ice

When Dave went downward in the Atlantic Ocean via the marine snow that dropped toward the deep ocean, 32,000 years ago, Icille (who had been at the surface 200 kilometers further north in a bicarbonate ion) entered the gas phase, flew up across the air-sea interface, and went back into the atmosphere as CO_2. Winds took her around the world twice. Then she was brought into living tissue by a sedge plant living in the border zone of ice-age climate between forest and tundra in what is now northern Germany. The sedge was eaten by a taiga antelope, which, during its southern migration a year later, was ambushed and consumed by a now-extinct species of cave lion. This lion, during an encounter with a local band of Cro-Magnons, became the inspiration for the artist who drew the lion that is still visible today on the walls of Chauvet Cave.

Now, we could easily say that the subsequent details of Icille's path were lost in the obscurity of time but that her next 32,000 years were statistically like Dave's described above. That would have been the case for most of the carbon atoms liberated along with Dave from the limestone cliff. However, Icille took a course that would eventually (close to our present time) win her carbon-atom fame almost equal to Dave's. But to reach this exalted level in the annals of science, Icille first had to become locked in ice.

The ice-age lion died in a ravine, and its carcass was consumed by a cave bear, by black vultures, and by beetles and soil bacteria. It was from the mouth of one of the vultures that Icille emerged again as airborne CO_2.

As was noted above, on average a carbon atom stays airborne as CO_2 for about 4 years. Roughly half of its exits from the atmosphere occur when it passes into a land plant, roughly half when it wriggles into the ocean. But occasionally, something extremely rare happens.

Five years after leaving the vulture's mouth, having been swept north, south, around the world a dozen times from the west in the mid latitudes and another dozen times from the east in the tropics, Icille found herself over Antarctica. As the breeze in which she cruised slowed down along the top of the 2-kilometer-high ice sheet that covered the continent (just as it does today), she came to a standstill, which allowed her to diffuse, from random molecular jitteriness, down into an air passage between crystal lattices of ice. She was still floating down in there a few decades later when the crack above her closed permanently as snow continued to fall and compress into ice. Icille could no longer leave. Like the other molecules in the bubble of air she was in, she was sealed in an icy microscopic chamber, which eventually would deliver a valuable cargo of ancient atmosphere to scientists.

The CO_2 of the Ancient Atmosphere

In the early 1990s, Dave and Oiliver had begun a period of lodging together in a cellulose molecule in the bark of a small, scrubby tree in Rwanda. It was then that Icille was liberated from the icy cage that had held her for 32,000 years while Dave had been circulating freely. Icille's freedom did not result from climate-induced melting of the ice that had trapped her. She was sprung by humans.

Paleoclimatologists had dug down into the ice cap at sites in Antarctica with equipment that looked like well-drilling rigs. Their goal was to pull up to the surface intact cylinders of ice from hundreds of meters down, and ultimately from a kilometer or more down. They then boxed and labeled sections of the ice cores, loaded them into special planes with freezers, and maintained the cores well below the freezing point all the way to laboratories in the United States and Europe. There the cores were sawed into disks like thick slices of salami. Eventually these were melted in highly controlled vacuum chambers so the bubbles of the ancient air could be extracted and analyzed.

One crucial measurement was of CO_2, which provided the paleoclimatologists with the value of that greenhouse gas in the atmosphere at any time, from hundreds to hundreds of thousands of years ago. It had been determined that these analyses did indeed yield the ancestral CO_2 values; in other words, nothing else had happened that could have altered the air bubbles' CO_2. The paleoclimatologists had also figured out how to date the ice as a function of depth by counting seasonal layers and by running mathematical models of ice compression over time. In addition, they had determined the age of air in the bubbles, which is always somewhat younger than the ice itself, with the offset depending on the rate of accumulation of snow and "sealing time" of the ice in the specific place where the ice core was taken.

The air bubbles in the ice were thus more useful than the usual fossils of ancient organisms, in which bones have been highly altered, many times into stone. The ancient air was no simulacrum but the actual material from the past. Getting it was perhaps somewhat like holding actual tissues of extinct creatures rather than just their traces in rock.

Eventually we will look at the ice bubble that included Icille. But here, let's go back in time, first in small steps and then in bigger ones.

At an Antarctic site called Law Dome, a giant mound of ice not far from the coast of Antarctica, paleoclimatologists were able to locate an ice core where the closure of the ice around the bubbles was so rapid (because of a high rate of snowfall) that air bubbles from it overlapped in time with the real-time sampling of the atmosphere provided by Mauna Loa.

Figure 6.1 shows the data from Law Dome[1] going back to 1840, along with the Mauna Loa annual average. The ice-core data are shown as dots, because they are from individual samples. In some cases, when several samples were from the same time zone, the dots are not exactly on top of one another, but they are reasonably close. The numbers are not as precise as the Mauna Loa data, of course; an ice core cannot resolve any seasonal cycle. But the ice-core data are remarkable and telling. With these preliminaries out of the way, the following facts can be read from figure 6.1.

First, ice-core data are available up to 1969. After that date, the layers of fallen snow still allow air to move in and out. But that date is good enough to overlap by 10 years with the real-time samples from Mauna Loa. The CO_2 values from the two data sources correspond closely over the course of that 10-year period, which supports the hypothesis that the trapped air bubbles preserve actual past air. (We need not worry that the ice-core air is from Antarctica, because we have

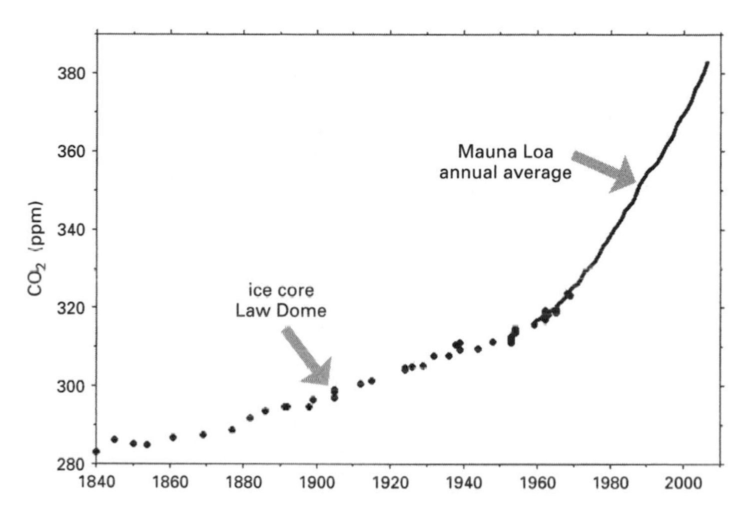

Figure 6.1 Law Dome CO_2 data (1840–1969) and Mauna Loa annual average (1959–2006).

already seen that the atmosphere is so well mixed that the air's CO_2 is basically the same everywhere for the annual average at the major monitoring stations, so long as they are not in urban areas.)

Second, the CO_2 level is lower the further back we go in time, at least to about 1840. At that date, about 20 years before the start of the Civil War, the air's CO_2 was just above 280 parts per million. That is 30–40 ppm less than when the Mauna Loa measurements began. This lower value seems to make sense, because humans were emitting CO_2 all during the twentieth century, all the way back to the middle of the nineteenth century (see figure 4.4), and even back to the start of the Industrial Revolution. However, one can show by comparing the fossil-fuel emissions against the rise in the atmosphere that the emissions are too small to account for the increase in CO_2 during the second half of the nineteenth century. Fossil fuel (coal back then) certainly contributed, but the main cause of the increase was net emissions from land-use changes, primarily from deforestation in North America. Not until around 1900 did the fossil-fuel emissions surpass those from shifts in the use of land. When these two sources are figured together, across time, the ice-core data confirm that humans have been altering the atmosphere for a long time.

Additionally, the rate of the rise apparently went up around the beginning of the Mauna Loa data. As can be seen in figure 4.4, the rate of fossil-fuel emissions of CO_2 accelerated around 1950, corresponding to the post-World War II scale-up of industrialization in many places.

But what about even further back? As was discussed in the preceding chapter, there are huge natural carbon fluxes in nature. For example, CO_2 shuttles back and forth between the atmosphere and the global ocean at annual rates exceeding 12 times the fossil-fuel emissions to the atmosphere. Photosynthesis by plants and respiration by microbes, fungi, and animals on land create CO_2 fluxes about as great as the air-ocean exchange. We now see the increase in CO_2 both in the direct measurements from sites around the world such as Mauna Loa, and also in the extended record back in time provided by the Law Dome ice core. Can we be sure that this increase is not due to natural imbalances in some of these largest fluxes of the global carbon cycle?

Measuring Back 1,000 Years

The ice cores allow us to look back even further. If the rise is attributable to humans, we would expect to see a period of time before the start of fossil-fuel emissions in which the CO_2 is not changing, or changing very little, if the fluxes in nature were in balance. But if we find substantial changes in CO_2 even before the Industrial Revolution, we will have to revise our expectations of how well the natural carbon cycle is bal-

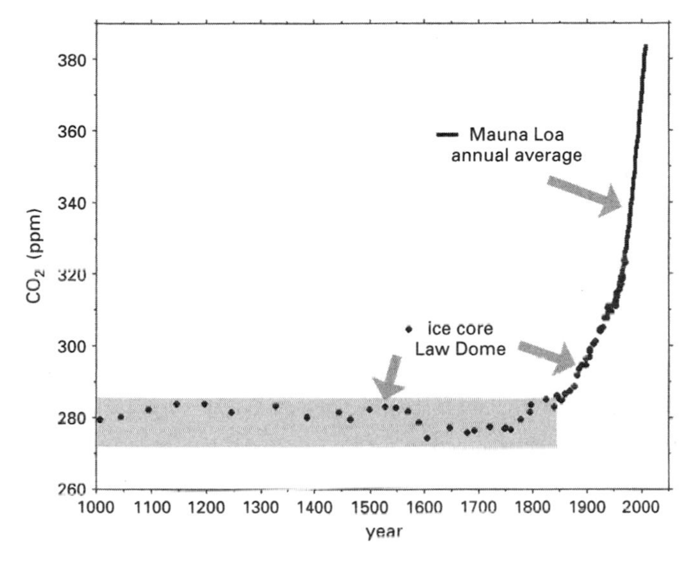

Figure 6.2 Law Dome CO_2 data (1000–1969) and Mauna Loa annual average (1959–2006).

anced. Fortunately, we have the numbers. The measurements from Law Dome go back 1,000 years. We can see what the CO_2 level was during the Battle of Hastings in 1066, or when Michelangelo was painting the Sistine Chapel in 1509. The data are displayed in figure 6.2.

Look at the data within the broad horizontal swath of gray from year 1000 to year 1850, which shows the points falling within a rather narrow envelope of constancy during that time. The individual dots do go up and down within the gray band, but by not much more than 10 ppm of total change, until around 1850, when the ice-core CO_2 climbs away from the long-term stable average of about 280 ppm. With CO_2 now climbing past 385 ppm, we can say with confidence that it now exceeds the pre-industrial average value by more than one-third.

Thus, the constancy of the atmospheric composition from year 1000 to year 1850 shows that pools and fluxes of the global carbon cycle must have been very nearly in balance. For example, the large annual fluxes that go from air to ocean and back must have been nearly equal, if not every year then at least for reasonably small groupings of years. The same conclusion holds for the large annual fluxes of carbon between the atmosphere, land plants, and the soil: the ins and outs across these pools must have been very nearly balanced.

Therefore, the natural levels of carbon in the biosphere's pools before the Industrial Revolution were quite constant (at least the atmosphere was, and because that pool is the main connecting link to all the others if they had changed so would it). Had the fluxes to or from the atmosphere been out of balance by even a billion tons a year (current industrial emissions exceed 8 billion tons a year), that would have led to changes over the 850 years of about 200 ppm. Obviously, nothing like that occurred.

Looking Back 12,000 years

What about even earlier? Ice cores from other Antarctic sites offer longer records. Here we must look not to Law Dome, with its fast snow accumulation and bubble closure but limited long-term record, but to Taylor Dome—a huge ridge, deeper within the continent, where Australian scientists have drilled. The ice record from Taylor Dome goes back 60,000 years. Here I will focus on only the most recent 12,000 years. Figure 6.3 shows the data.[2]

The date of 10,000 B.C.E. on the left side of the graph takes us "just" before the beginning of agriculture. The earliest great Egyptian pyramids of Giza were built around 3000 B.C.E. (For dates before civilizations started keeping dates, scientists usually use B.P. for "Before the Present," but I'm going to stick with B.C.E. for "Before the Common

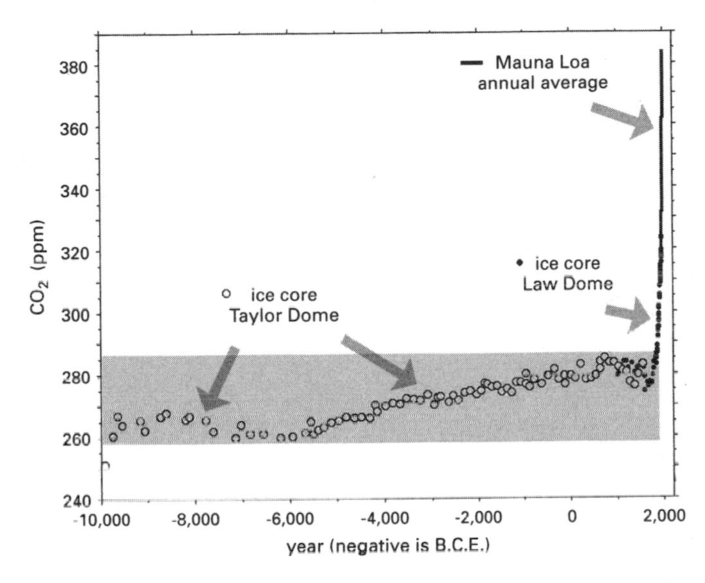

Figure 6.3 Ice-core CO_2 data from Taylor Dome in Antarctica, going back 12,000 years, and Mauna Loa annual average (1959–2006). The data from Law Dome are also shown, to provide a continuous record from the final date of Taylor Dome (1560) and the first date of Mauna Loa (around 1960). The gray bar emphasizes the long interval of relative constancy.

Era," also known as B.C., and C.E. for "Common Era," also known as A.D.) Again, a gray bar emphasizes the steadiness of the CO_2 over many thousands of years. We do find within the gray band a gradual long-term upward trend, but it's not much. Twelve thousand years ago the atmosphere's CO_2 was around 260 ppm, and it rose by about 20 ppm over the next 10,000 years. That's an average increase of about one ppm every 500 years. Right now the rise is close to two ppm per year. So the current rise is taking place at a rate between 500 and 1,000 times the rate of the "natural" rise that occurred during those 10,000 years.

The cause of the small rise over 10,000 years is still being debated by scientists who want to decipher the natural workings of the carbon cycle because it will help them understand more about the ways that putative shifts in the dynamics of the ocean and land pools can affect CO_2 levels—knowledge that we need in order to predict the future. But the main finding from these data is surely that until about 1850 the fluxes of CO_2 must have been very nearly in balance, century after century, for thousands of years, with any small excursions cancelled by opposing ones, like a seesaw nearly balanced by wriggling a bit up and down.

I have made the graph in figure 6.3 start about 12,000 years ago for a reason. That is about the time when the last glacial age ended. At the farthest left-hand section of data, one lone dot at 10,000 B.C.E. is well below the gray band. This is a portent of what will be found when we go back even further in time with CO_2 from ice cores.

Icille's Ice-Age Message

Now let us consider the data from the ice that held Icille. Data from the Taylor Dome ice core go back that far. That ice was brought up in the early 1990s, and after its analysis Icille was released back to the atmosphere. Her presence in the ice helped provide the data everyone can now see in figure 6.4 for the atmosphere's CO_2 way back to 40,000 B.C.E.

The data show something remarkable. Right around 12,000 years ago (10,000 B.C.E.) the CO_2 drops, reaching a low of approximately 190 ppm about 20,000 years ago. There were some additional changes even further back, but they were minor. When Dave and Icille emerged from the limestone cliff, 32,000 years ago, the CO_2 was on the order of 200 ppm. So the natural carbon cycle has not been perfectly constant.

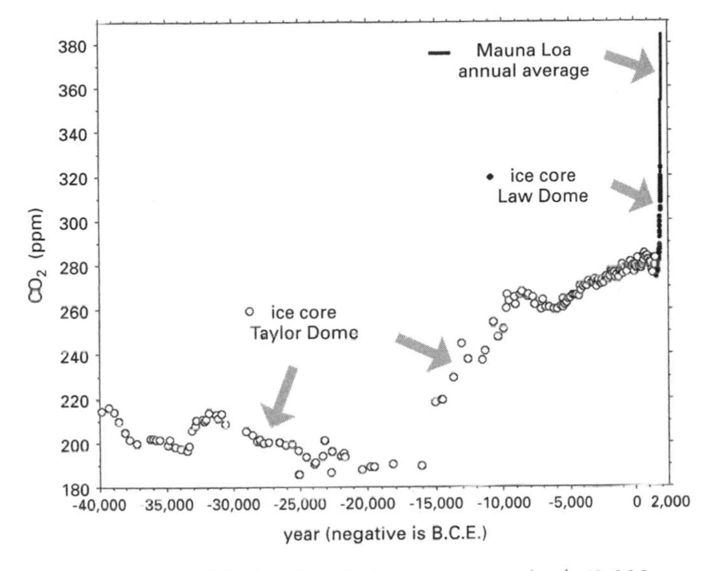

Figure 6.4 Ice-core CO_2 data from Taylor Dome, going back 42,000 years. The data from Law Dome are also shown, to provide a continuous record linking Taylor Dome to Mauna Loa.

These numbers take on more meaning when we consider the fact that the time of lowest CO_2 corresponded to the time of deepest glacial cold in the most recent ice age. The rise of CO_2 from 20,000 to 10,000 years ago corresponded to the transition out of the ice age into an interglacial (a warmer period between long icy intervals). So during the warm interglacial the CO_2 was between 260 and 280 ppm (before the start of the Industrial Revolution), whereas during the cold glacial era it was closer to 200 ppm. When Icille became locked in the ice, a few years after she and Dave were dissolved from the limestone cliff of Paleolithic France and the cave lion was being painted on the wall of the nearby Chauvet cave, the world was about 9°F (5°C) colder.

This finding blew many minds in the scientific community. A time of global chill had a reduced amount of the greenhouse gas. Could these facts be related? If the greenhouse gas affects climate, then during a time of lower CO_2 we should expect to find lower temperatures. Apparently we do. Ancient temperatures are obtained from estimates of the extent of former ice sheets, from fossil evidence, from models, and from oxygen isotopes, available from water molecules in existing ice. Indeed, using "predictive" models to calculate the cooling produced by the lower amount of CO_2 that led to a reduced greenhouse effect during the Ice Age, most paleoclimatologists accept the conclusion that the lower CO_2 was responsible for about half of the chill of the Ice Age. The other half of the cooling was due to the large extents of the ice sheets. Primarily sitting atop vast areas of land in the northern hemisphere, the ice sheets cooled the planet by reflecting more sunlight to space than during the warmer interglacial times—sunlight that would otherwise have been absorbed by Earth's more extensive tree-covered surface in the warmer intervals.

About 20,000 years ago, CO_2 began to rise as the Ice Age began to end, and the massive ice sheets in North America and Eurasia gradually melted. Twelve thousand years ago the CO_2 reached a level at which it remained until about 1850 (except for the gradual 20 ppm unexplained rise during that time). Paleoclimatologists have not yet explained in detail what caused the more dramatic rise in CO_2 as the Ice Age ended. But it is accepted that the cause must have been a change in the oceans, for only a carbon pool that large could alter the balance of the atmospheric pool and hold that balance in a new state over such a long period, without considering even longer-scale geological or evolutionary changes.

Interesting! Shifts can take place in the natural carbon cycle. If the post-glacial rise had to do with the ocean, could the modern rise also be partially caused by the ocean? I ask this question even though the modern rise is much more rapid, occurring over the interval of about two ink pixels in figure 6.4.

Carbon-cycle scientists harvest data from tree rings to measure the amount of radioactive carbon (carbon-14) relative to the more abundant stable carbon isotope (carbon-12). Carbon-14 is formed by cosmic rays hitting nitrogen in the air, and this radiocarbon, with a atomic decay half-life of about 7,000 years, gets incorporated into plants and therefore into the woody tissue of trees just like stable carbon. But then it decays back to nitrogen at the known half-life rate. The amount of radiocarbon in a tree sample can be used to date ancient building projects that used wood, such as the ceiling beams of the cliff dwellings in the ruins of Mesa Verde in southwestern Colorado.

Before the nuclear bomb tests of the 1950s, which contaminated the air with non-natural radiocarbon, tree rings revealed a decline in radiocarbon, say from 1900 to 1940[3]. This decline could only have resulted from the addition of CO_2 derived from fossil-fuel combustion, which has, essentially, zero radiocarbon because the fossil fuels are so old that all the radiocarbon has atomically decayed that was originally in them when they began as ancient plants and algae. This evidence shows that the modern rise is due to the combustion of fossil fuels, not to any lessening in the ability of the ocean to store CO_2, though it was this lessening that caused the post-glacial rise of atmospheric CO_2, which took thousands of years.

Nevertheless, the fact that natural changes occurred in past CO_2 levels shows that the balance in the amounts and fluxes of carbon can shift among the pools of the biosphere. This conclusion can be brought home even more dramatically by looking at the CO_2 data for

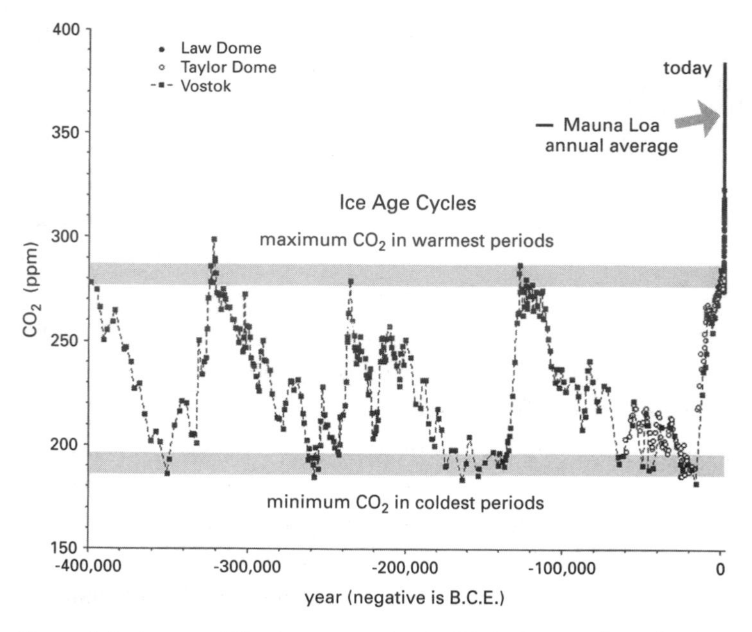

Figure 6.5 Ice-core CO_2 data, going back 400,000 years, from the Antarctic site of Vostok. For continuity and comparison, all the data from Taylor Dome, Law Dome, and Mauna Loa are also shown.

400,000 years[4], which became available from another Antarctic site called Vostok. Named after the Russian word for east, Vostok is located very near the South Magnetic Pole. It holds the record for the coldest temperature ever recorded on Earth (–89.2°C, –128°F). The low annual rate of snowfall at Vostok has created a deep record of ice. In the 1980s, an international team of paleoclimatologists made headlines by pulling up and analyzing a long sequence of ice cores. The findings are illustrated in figure 6.5.

Over the entire interval of 400,000 years, the records of CO_2, temperature, and volume of the ice sheets show that there were four ice ages, with a cycle for of about 100,000 years. The causes of the ice ages are still being debated, but they definitely correspond to various periodicities in Earth's orbit around the sun, and they indicate that Earth's climate can be extremely sensitive to small changes in the amount and seasonal distribution of the solar energy received by the surface. And not only is climate apparently extremely sensitive; so are the biogeochemical cycles, for, as we see in the Vostok record, the CO_2 content of the air has shifted up and down with the retreats and expansions of the ice sheets, during highs and lows in global temperature. Note that at no time except for a few slightly higher points during the last 400,000 years has the atmosphere held more than about 280 ppm of CO_2, which is about 100 ppm below the current value of more than 380.

More recently, data from yet another site (Dome C in Antarctica) have yielded air bubbles 650,000 years old.[5] This time span includes two additional grand ice-age cycles, and in these cycles the CO_2 is also high during the warm periods and low during the cold periods, just as at Vostok. The range of the CO_2 during the two additional ancient cycles revealed at Dome C is not quite as large as its range during the more recent four cycles, a fact that provides more intriguing grist for the mill of understanding for the paleoclimatologists. But the overall picture is consistent: At no time in the last 650,000 years has the atmosphere's CO_2 been anywhere near what it is today. It has gone as low as 180 ppm, creating a cyclic pattern of rises and falls of about 100 ppm in amplitude in sawtooth waves with cycles of roughly 100,000 years. Thus, the CO_2 values had a remarkably constant maximum and minimum. Today's atmosphere is more than 100 ppm above that maximum. This rise is taking place over an interval that is a small fraction of a dot of ink in figure 6.5. Fossil-fuel combustion has created an atmosphere that contains nearly 40 percent more carbon dioxide than during previous warm periods along an interval of 650,000 years.

7

Wealth, Energy, and CO_2

Today the rising level of CO_2 is markedly higher than it was at any time than during the biosphere's past half-million years. Natural cycles did occur in which the distribution of the biosphere's carbon shifted, and these coincided with ice-age cycles. But those past shifts were slower than the rapid one that we are now in. More recently than these ice-age cycles, for the huge stretch of time during the past 12,000 years up until about the year 1850, the carbon cycle was very nearly in steady balance. Without doubt, humans have put the atmosphere into a new chemical state.

How high will the CO_2 concentration go? There is an old joke about government waste: a billion dollars here, a billion there, and pretty soon we're talking about real money. The same goes for CO_2 as shown by the Mauna Loa record: a billion tons here, a billion there, and soon we're talking about real shifts in the greenhouse atmosphere. Given current trends, we can project the future of emissions.

Although forecasting the future of something as complex as global industrial civilization is fraught with uncertainty, there is less uncertainty here than in, say, forecasting the course of love after a first date. We might, for instance, simply extrapolate the current rate of the CO_2 rise. We might also try to unravel some of the dynamics. A person running down the street and exhaling heavily could be jogging for pleasure or could be running to buy milk before the store closes. An intention is in play. Moreover, it might be important to ask how the runner's legs work as part of the body's musculature, and whether those legs are nearly spent or still strong. Though the global industrial system does not have intention as a collective, the individuals within it have desires and use their time and effort toward particular ends. And though the global industrial system does not have biological muscles, there are internal connections that make the system run. Let us look at some of the large-scale directions and internal dynamics of the global industrial system that are relevant to projecting CO_2 emissions.

Oiliver and Methaniel Help to Cook a Meal in Rwanda

Year 2004: In a hut in a village in the impoverished and recently war-ravaged central African country of Rwanda, Oiliver and Methaniel are inside a branch that had been harvested from an acacia tree. The branch is being fed into a large fire, the heat from which will cook a meal eagerly awaited by children in a school (figure 7.1). It is the middle of the official school day. The children rely on at least one good meal, because what will await them at home is not certain.

By a twist of carbon-cycle fate, our two atoms have found themselves not only in the same species but in the same plant. They are chemically bonded as parts of a chain of carbon atoms in the cellulose molecule (see figure 1.4), the most abundant part of woody plant tissue. Cellulose may not be good to eat (unless you are a termite), but it's very good for burning.

Here there is no high-tech processing such as has been proposed by modern energy engineers who want to convert cellulose to liquid ethanol for biofuels. The Rwandan cook simply lays the stalks on the

Figure 7.1 A meal being cooked at a primary school in Rwanda. Oiliver and Methaniel, neighbors in a cellulose molecule inside a stick of combustible biomass, emerge in two separate molecules of airborne CO_2 (upper right), which go up the chimney.

floor and slowly feeds them into the flames. The top portions of the branch have already been fed in. As Oiliver and Methaniel approach the flames, they are torn apart. What had been democratically shared electrons between the two are grabbed during the combustion by oxygen atoms from the air, which assemble new, asymmetric bonds with the carbon atoms during the burn, releasing as heat and light a portion of the electrical energy formerly possessed by Oiliver and Methaniel, and creating as waste many small molecules of gaseous CO_2 from what had been large, solid, sprawling cellulose molecules. The rising current of smoky, hot air takes the small waste molecules out of the hut and back into the atmosphere.

Ten years earlier, about the time when Icille had been liberated from the ice core, Oiliver and Methaniel had separately, each as airborne CO_2, entered the leaves of the growing acacia tree, and as a result of the tree's photosynthetic biomolecular machinery had been shunted together into a brand-new cellulose molecule. To burn the stalks seems a perfectly good example of a sustainable form of energy. And, ideally, it is: the tree grows for a number of years, is cut, and then is burned. Initially, carbon goes from the air into plants. When plants are burned, there is no net release of carbon, just a return to the air of the same carbon that was taken from the air some years before. We saw such a sustainable cycle in the way that the beer's carbon in alcohol, which came from the air, became CO_2 in transformations involving the drinker's bloodstream and lungs, and was then exhaled.

Thus, in principle a tree's growth and its combustion can form a sustainable cycle as long as the rate of combustion (say, in a village or a country) is not faster than the re-growth of the fuel; otherwise, the result will decrease the standing biomass. In that case, there is net deforestation—and we know that often is the case.

Rwanda, in using wood for fuel, is not a model of economic development achieved using sustainable energy. It is just the opposite. Countries in which biomass is regularly used for cooking tend to be poor. They are what economists call "developing" or "undeveloped." Combustible biomass is often all there is to cook with. And often it is biomass of poor quality, such as crop residue or dung. It is used because it is available "free"—it can be locally harvested by families for their own use, or by villagers who can sell it in the open market. Biomass-burning stoves are dirty, fumy, and difficult to regulate, unlike gas or electric stoves. And land is required to grow the biomass.

Rwanda does emit CO_2 from fossil fuels. To put this flux in perspective, the average citizen of the United States emits as much fossil-fuel

CO_2 in a day as an average Rwandan emits in 10 months. Rwandans are poor because they are not using fossil fuels and generating lots of CO_2 emissions. Presumably they want to burn more fossil fuels in the future, and they probably will. Otherwise, it seems, continued poverty is their future.

Rwanda is a small country with only 8 million people, so what the Rwandans do in terms of CO_2 emissions will not make that much difference to the global greenhouse levels. But larger countries in a similar state of nascent fossil-fuel development could affect the future substantially.

In Bangladesh, Coalleen Feeds a Fire and Dave Feeds a Body

One of the most prominent futurists of the twentieth century was the inventor and architect R. Buckminster Fuller. He often emphasized how ordinary citizens of the developed countries live in conditions of relatively enormous wealth, equivalent to the lifestyles enjoyed only by the nobility hundreds of years ago, all because a billion moderns have large numbers of "energy servants" in the forms of fossil fuels.[1] In contrast, ancient nobility relied on the muscle energy of people and animals working for them as poorly paid or unpaid servants.

Any external source of energy that helps support human life could be considered a kind of energy servant. The combustion of biomass to cook food goes back at least hundreds of thousands of years and possibly, according to some anthropologists, a million years or more.

Decades ago I once relied on an energy servant while on a camping trip to a Canadian lake. I had become enamored of the fact that dried moose dung was, in essence, combustible carbon just like wood or any fossil fuel. My companions were furious when I ruined the taste of a fresh-caught Northern pike, one of the most delicious freshwater fish.

Today, in rural Bangladesh, dung patties are dried for household use by spreading them out on surfaces such as trunks of steeply leaning trees. The households have side sheds where every manner of combustible biomass is sorted and stored: the dung patties, branches of trees, leaves, and, most resourcefully, inedible stalks left after the harvests of field crops such as rice.

The same day Oiliver and Methaniel were released as CO_2 from the fire in Rwanda, a dinner was prepared in one of those rural villages in Bangladesh. There, Dave and Coalleen had also been about to partake, in a poetic sense, of a meal. From the air, they had both entered a rice plant during its previous growing season, which meant that they had had 10 years of other adventures in the biosphere while Oiliver and Methaniel had been locked in the branch of the growing acacia tree in Rwanda. Coalleen was in lignin (a tough, tar-like molecule) in a rice stalk that was being inserted into the fuel hole of the clay stove. Dave was in a starch molecule inside a grain of rice that was being cooked with the help of Coalleen and her neighbors in the stalk. Both Dave and Coalleen were to end up as CO_2 molecules released back into the air (figure 7.2). In the stove's combustion chamber, Coalleen was immediately burned with oxygen, as Oiliver and Methaniel were burned in their fire, yielding CO_2. Dave was eaten, digested in the eater's small intestine, and turned into CO_2 inside one of the mitochondria inside the muscle cell of the eater's left thigh when she walked out to tend the rice field the next morning. The thermal combustion process of burning biomass produces CO_2 and energy. So, too, the body "burns" complex carbon molecules that are eaten, though at the much lower body temperatures, and produces CO_2 and energy. The overall chemical reactions in the flame and in the body, in their simplest forms, are identical:

$$\text{Any organic form of carbon} + \text{oxygen} \rightarrow \text{freed energy} + CO_2.$$

Figure 7.2 A meal being cooked in rural Bangladesh. Coalleen, now in a lignin molecule inside the combustible biomass (in this case, stalks of rice plants), will emerge from the stove's vent in a CO_2 molecule. Dave is in the rice about to be cooked in the pot and will be exhaled as CO_2.

At a fundamental level of analysis, when agriculture relies on human power (or animal power) the energy in the food produced has to be much greater than the energy used to grow the food, which is the sum of the muscle energy it took to till the soil, plant the seeds, weed, irrigate, harvest, and process, because that food is the sole source of the energy for the human and animal labor. Modern societies use external fossil-fuel energy servants to make their systems of agriculture more efficient in terms of reducing human labor, which is why in those societies the agricultural labor sector might require less than 1 percent of the total working population. But the energy put into producing food may then be many times the energy value of the food. In Bangladesh, in contrast, without large amounts of fossil-fuel energy, even rice stalks are a valuable source of energy.

Because the combustion of fossil fuels and the human body's energy-creating metabolism both release CO_2, one way to compute the amount of fossil-fuel energy servants is to compare the amount of CO_2 produced in fossil-fuel energy use by an average person of a country (or a region) against the amount of CO_2 produced by the average human body. The fossil-fuel CO_2 released per year by the average Bangladeshi was about 70 kilograms of carbon in 2004, the most recent year for official data from the Carbon Dioxide Information and Analysis Center. The carbon in the food in an average human's diet amounts to about 90 kilograms per year, most of which will be released as CO_2 in breathing and the rest as liquid or solid waste, which will fairly quickly be turned into CO_2 by bacteria. So in Bangladesh, the per capita release of CO_2 derived from fossil fuels is less than the release of CO_2 from one human being (70/90 = 80 percent). A little less than one fossil-fuel energy servant is cranking away outside the body of each person, augmenting their bodily energy just a little.

What will happen as Bangladesh develops, inevitably using increasing amounts of fossil fuels? Bangladesh has more than 140 million people, though its land area is only about the size of the US state of Georgia or about one-third the size of France. If Bangladesh industrializes to the level of the United States in terms of energy use per person, and if it creates that energy by combusting fossil fuels, then the world will have added a "region of emissions" equivalent to nearly half of the United States. Poverty versus wealth as related to energy is more than an economic issue involving available capital and the marketplace of supply and demand; the issue is integral to the topic of future atmospheric CO_2. As poorer countries become wealthier, their fossil-fuel CO_2 emissions almost invariably increase. Issues of social justice aside, we must look at the related topics of economic well-being, energy, and CO_2 emissions to understand the driving forces behind the trends in atmospheric CO_2 and to make a reasonable prediction about how high CO_2 will go in the future.

An Average Person: The United States versus the World

I have been using the United States as a standard to entice you to entertain the notion of a world in which countries become more like the United States in terms of their citizens' CO_2 emissions. Why the United States? Well, until 2006, when China surpassed it as the biggest emitter,[2] the U.S. had held that position for decades. However, the crucial value is not the emissions by any country (important as that is) but the per capita emissions, and for that too the U.S. has been way up there. Of the relatively populous countries, say over 100 million, the U.S. has led for decades and still leads in terms per capita emissions. The United U.S. has set the example, providing the world with an implicit goal for economic growth: To get wealthy, burn fossil fuels. It might be an instructive fantasy to compute a future in the opposite direction, say, what the world's emissions would be if they were the same on a global per capita basis as perhaps China or India. But the real lesson comes from using the U.S. as the exemplar.

Why is the per capita basis so important?

For some considerations, countries are the natural units to use. Countries create internal laws, such as pollution standards and taxes on fuels, and agree or not to treaties with other countries. But for analyzing a pollutant to which every person contributes by participating in their country's economy (including children being raised), comparing two different countries by their total emissions is a classic situation of comparing apple and oranges, because countries can vary so much in their populations. If it were merely the totals for countries or regions that were important, then the U.S. could say "Look, we as a nation emit one-fifth of the world's CO_2. But all the world taken together (as a unit) emits four-fifths. So it's everyone else that is most of the problem." Technically the numbers are true, but this "logic," which gets used when a U.S. politician wants to dismiss national responsibility (yes, it has happened), neglects the fact that the U.S. has less than 5 percent of the world's population and has by example set the paradigm to which many other countries aspire. That paradigm, in which wealth and CO_2 emissions are related, can only be analyzed properly across the world's countries by focusing on the per capita numbers. So just how closely are wealth and emissions related?

Using data from the Netherlands Environmental Assessment Agency for 2006, the fossil-fuel CO_2 emitted by all the activities of the United States was 1.6 billion tons of carbon. This number is the sum of emissions from all sources, including automobiles, fossil-fuel power plants, the use of fossil fuels by homes, commercial sites, and industry. The U.S. population in that same year was 301 million, so we can compute the per capita emissions: 5.3 tons of carbon per U.S. citizen, which went up into the atmosphere in 2006 as fossil-fuel waste CO_2. In terms of fossil-fuel energy servants per person (using the human body's metabolic CO_2 release cited above, and recalling that for an average Bangladesh citizen the number was less than one fossil-fuel energy servant, the U.S. had nearly 60 fossil-fuel energy servants working for each citizen. One can argue that this is the foundation of American wealth.

Now for the world number. The world emitted 8 billion tons of carbon as CO_2 from fossil-fuel combustion in 2006 (a small amount in this number coming directly from the rocks used in cement manufacture, folded in as industrial emissions). The world's population was then 6.5 billion, and thus the world's per capita emissions of CO_2 was

1.2 tons of carbon. In terms of fossil-fuel energy servants, the world average citizen has the "respectable" number of 13 human metabolic equivalents at work in the moment-to-moment conversions of Oilivers, Coalleens, and Methaniels from fossil organic carbon into CO_2.

When the per capita emissions (tons of carbon as CO_2) of the United States are compared against those of the world, the ratio is 5.3/1.2 = 4.4. Were the world, therefore, to become everywhere like the United States, global emissions would increase to 4.4 times what they now are. That would mean even more serious fluxes of billions of tons of carbon, which would be worth not only talking about but worrying about.

Per Capita GDP, Energy, and CO_2 Emissions Compared

To look at the relationship between wealth and CO_2 emissions more precisely, we need a metric for wealth. The metric now used by the majority of economists is gross domestic product (GDP)—a measure of total economic output as the sum of the final values of all goods and services: the milk you buy, the steel sold to a construction site, the gasoline pumped into a car, the salary of a salesperson, the hydroelectricity sold by a power company, the fee a doctor charges for an examination, and so on. That many transactions in the developing world are not monetized is an issue, and the use of GDP as principle metric is an even larger issue. However, that metric is available as data for the world's countries, and it is the one used by the CO_2 analysts in the Intergovernmental Panel on Climate Change. Also, as anyone can tell from a comparison of the state of affairs in countries with high versus low per capita GDPs, the metric does work to at least roughly reflect the material well-being of a society. Gross domestic product is usually in units of U.S. dollars pegged to a particular year (to account for inflation), and is adjusted for individual countries by a factor called "purchasing power parity," creating some "play" in the numbers that economists debate but which do not affect the overall, first-order, big-picture patterns to which I will point.

To complete a link between CO_2 emissions and wealth, we also must look at energy. In our logic, energy is the muscle that powers the activity of the world economic "body," which is exhaling the waste gas. Not all forms of energy conversion emit CO_2, and energy can be used more efficiently or less efficiently. The numbers here will not include biological-muscle forms of energy (and thus do not include the energy of human or animal labor), but will include all other forms, whether they emit CO_2 or not (so nuclear and hydroelectricity are included). Numbers for energy on the scale of big countries or the world comes in a variety of units in the literature, such as quadrillions of BTUs ("quads"), million of barrels of oil equivalent, or terawatts (trillions of watts, technically a unit of power), and with care they can be interconverted. For present purposes the units are not important, because I will be citing ratios of different countries' total energy use to the world average (both on a per capita basis, of course). Note that this number is not just for a person's direct energy consumption at home or in a car; it includes all the energy that went into all purchased goods and services. The way I calculate per capita energy use is similar to way I calculated CO_2 emissions above: I simply divide the country's total energy consumption by its population.

The "bottom-line" goal is to look at per capita values for GDP, energy, and CO_2 emissions. Figure 7.3 illustrates these values for the United States relative to the world.[3] (For the world, the acronym is not GDP but GWP, for gross world product, but it is only the scale that is different, not its computation as an economic metric for wealth.) GDP, energy consumption, and CO_2 emissions are beautifully, if I may say

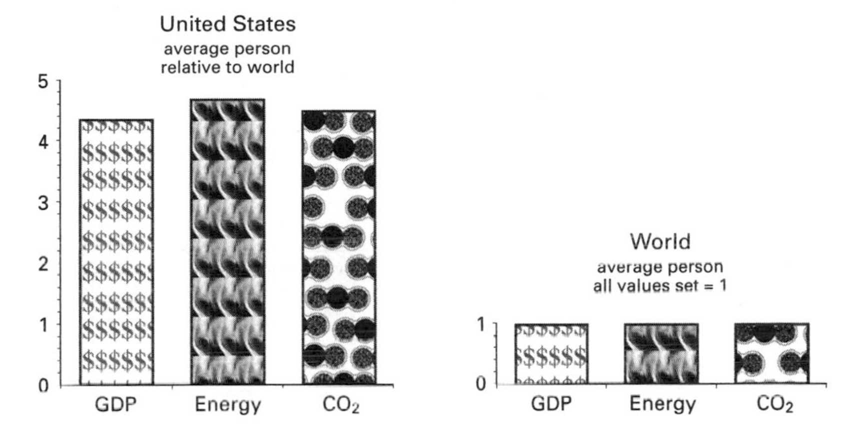

Figure 7.3 Data for the United States and the world in 2006: GDP (technically, GWP for the world), energy consumption, and fossil-fuel CO_2 emissions

so, correlated. For the United States, all three properties are 4.3–4.7 times the world averages.

In today's world, wealth on a large scale is (though with some variance) closely proportional to energy consumption, and that, in turn, is closely proportional to CO_2 emissions, because the dominating energy source is the suite of fossil fuels. You could almost say "Give me the number for the fossil-fuel waste and I'll tell you how wealthy the country is." In today's world, if you want to get wealthy beyond even the dreams of ancient kings and queens (certainly in terms of purchased goods, ease of travel, and much else, if not stolen gold), you burn fossil fuels.

Does the pattern hold when we look at other countries? In figure 7.4 we can see the bar-chart heights of the economies, energy flows, and CO_2 emissions for the European Union and Japan, which I have combined because their numbers differ by less than 10 percent.[4] The figure also shows data for China, for India, and for Africa as a unit.

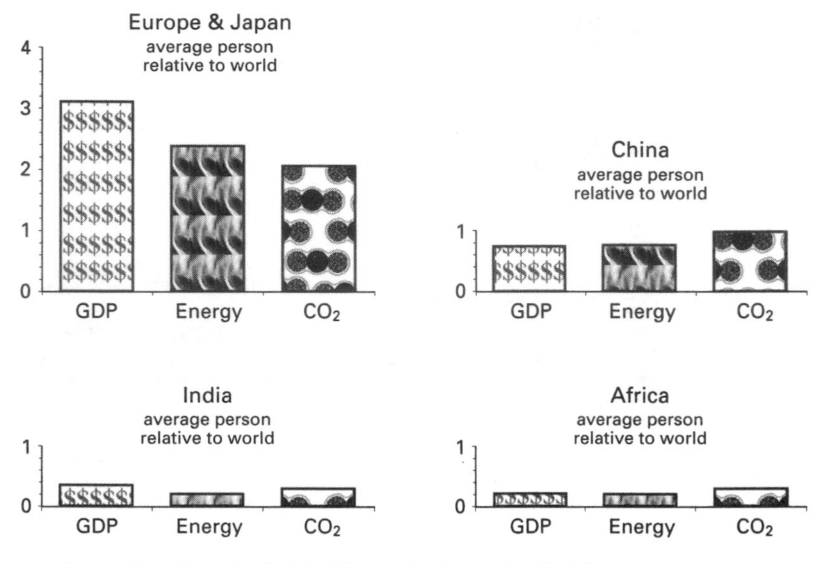

Figure 7.4 Data for 2006. "Europe" refers to the EU 15.

In general, our rule of thumb holds up just fine. India's GDP is extremely low, and, correspondingly, so are its energy consumption and its CO_2 emissions. The same goes for Africa. For China, all three parameters slot in at about the same, relatively intermediate levels. Interestingly, China's per capita fossil-fuel CO_2 emissions in 2006 were exactly the world average.

The European Union and Japan confirm again the rule that wealth is generated by lots of energy, which in turn, because energy means primarily fossil fuels, generates lots of CO_2 waste. But there is one difference between the picture for the EU and Japan and that for the United States, neglecting the minor fact that the per capita GDP for the EU and Japan is somewhat lower than that of the United States. (They are all up there in the club of highly developed countries.)

Whereas for the United States the bars for energy and CO_2 are both somewhat higher than the bar for GDP, for the EU and Japan the bar for energy is lower than the GDP bar, and the bar for CO_2 is lower still. So although the EU and Japan fit the general pattern (that is, all the bars are high relative to, say, the depressed bars of India and Africa), the specific heights of the three bars also tell a story. And for the EU and Japan that story is that units of GDP are generated with fewer units of energy than in the United States, and the energy used generates less CO_2. Thus, both the EU and Japan are more energy efficient at generating wealth.

Some of the difference that Europe and Japan share in comparison to the United States has to do with geography: they are denser, so transportation distances are shorter, and people drive relatively small cars and make more use of trains. Also, people often live in more compact dwelling units, such as small homes or apartments with shared walls that reduce the heating and cooling requirements. All these features reduce energy use. Here I only want to emphasize that the details of the developed infrastructure of a country or a region can make a difference. It is thus not a law of development that GDP, energy, and CO_2 must always correlate *perfectly*. A rough correlation is apparently the general rule for the present state of the world, but the correlation is not a law of the universe. In fact, to design a future in which the global society emits less CO_2, we must simultaneously bring down the energy bar relative to the GDP bar, using energy conservation and efficiency, and also bring down the CO_2 bar (even more so, relative to the energy bar) through large-scale deployment of sources of energy that do not emit CO_2. Relative to the U.S. pattern, one reason the CO_2 bar of Europe and Japan is lower than the energy bar is that those regions generate a higher fraction of their electricity by nuclear fission. Brazil is another country with a proportionally low fossil-fuel CO_2 bar.

That is attributable to Brazil's substantial and growing production of ethanol liquid fuel from sugar cane, which is sustainable (considered only from a carbon point of view) because the sugar cane takes carbon from the atmosphere and then Brazil produces a carbon fuel from the sugar cane to burn, which returns the carbon to the atmosphere. This process is thus not a net generator of CO_2 in the way that fossil-fuel combustion is.

The Global Industrial Growth Automaton (GIGA)

Predicting the future growth rates of CO_2 would be complicated if valid projections were required for the economic futures of all the countries of the world. And their individual economic paths have varied a lot. China's blistering progress is now in the news daily. Japan once had a growth rate as eye-catching as plum blossoms opening in spring but has since mellowed. Vietnam is rising with remarkable vigor. Russia endured rough times, and actually had an economic contraction in the 1990s. Sadly, Africa has suffered negative growth, in terms of constant economic dollars, over the last several decades. How should we project the courses of all these different countries and regions, with their unique pasts and their presumably unique futures?

It is worthwhile taking a big-picture look at the world as a whole, say in the way an ecologist might be concerned about the growth rate of a forest irrespective of the varying rates of the different tree species within it. I hope you might admire the results of such a broad-brush analysis, because the resulting graph, at least to my eyes, appears as significant as, say, the Mauna Loa rise in CO_2 or the historical data from ice cores.

In figure 7.5 I plot the gross world product in constant 2005 dollars, using the natural logarithms of the numbers.[5] If this semi-log plot

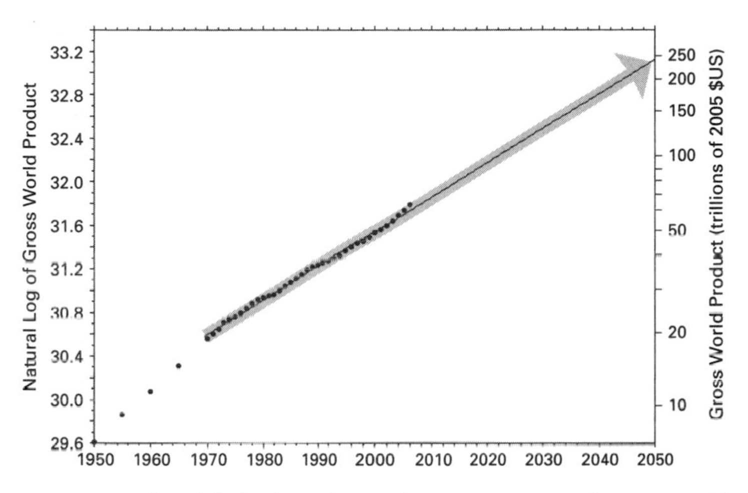

Figure 7.5 The Global Industrial Growth Automaton. Data for gross world product from about 1970 to 2006 show a remarkably persistent 3.1 percent annual growth rate. (A straight line going up on a natural-log graph shows exponential growth, with the slope representing the rate.) Here I project this to 2050. It is interesting that the earlier data, from 1950 to 1970, indicate that the growth rate during that time interval was higher—about 4.6 percent annually.

reveals that the data lie along a straight line, we can determine that there is an exponential growth rate in the world economic system, and we can calculate a numerical value for that exponential rate by the simple slope of the line. (The right side of the graph gives the actual numbers for the gross world product in trillions of dollars. The intervals of that scale are not, of course, uniform, because the data graphed have been converted to log numbers, as noted.)

The data after 1970 can indeed be fitted with a line with a very high statistical metric for goodness of fit. The slope of this best-fit line is 0.031, which translates into an annual growth rate of 3.1 percent. Note that the growth rate before 1970 was even higher (points are only shown every 5 years), at about 4.6 percent annually between 1950 and 1970, due to the post-World War II boom in certain countries that dominated the world economic scene. But now the world is more integrated and more complex, and it is chugging along with a 3.1 percent per year increase in total economic activity. Yes, there have been slight undulations in the data, quivering spurts and slowdowns, but overall the 3.1 percent growth line has been astoundingly steady for more than three decades.

I like to call this global system the Global Industrial Growth Automaton—GIGA for short. Why? Obviously it is global, industrial (including services associated with our industrialized age), and growing. The "automaton" part may be less obvious. Consider that no one is conducting the GIGA along its path. The course comes from the collective result of the activities of what are now nearly 7 billion individuals pursuing their needs and dreams within the contexts of more than 200 various home countries. The GIGA chugs on like a machine on a track.

Or perhaps the GIGA is dynamically more akin to another process with exponential growth: a biological population during its stage of expansion. Think rabbits in Australia, or springtime algae in a pond, or bacteria in a Petri dish. In an expanding population, exponential growth is a result of a constant unit reproductive rate applied to the expanding population. If one tree puts out a seed that will germinate and grow into an adult plant every fifth year, the annual growth rate is 20 percent. A group of 100 trees still has a growth rate of 20 percent, but the total growth is 100 times that of one tree. Thus, a constant unit rate leads to an exponential increase in the overall numbers—the same universal magic that economists call compound interest.

In exponential growth, more engenders more. The new rabbits, algae, and bacteria become producers themselves. A growth rate for the GIGA is grounded in the fact that some of the results of economic growth each year go into machinery for obtaining more energy and materials—for example, for producing more rigs for drilling for natural gas and oil. Put broadly, we not only manufacture; we manufacture the devices for more manufacturing. Profits from development are plowed back into further development. Thus, the parts of the industrial system can be said in a sense to be reproducing. For particular businesses, fortunes rise or fall, as individual rabbits that might thrive or die. But from a larger perspective the fates of individual businesses, rabbits, or bacteria are statistical noise in the overall growth of the population as an entity. In the case of the technological world, this entity is an increasingly integrated population of GWP-producing laborers and their machines.

Over the course of time represented in figure 7.5, the dynamics of the GIGA have transcended countries, local wars, political regimes, the foibles of leaders, types of technologies, and even the weather, and these dynamics embrace the complex connections among industrial cities, factories, roads, malls, high-rise buildings, and slums. It might not always feel so great being a cog in the great self-replicating, biology-like, industrial world's fossil-fuel machinery, and that does seem to be the message about who we are from the graph. But for those in industrialized countries with fossil-fuel energy servants—such as all the passengers on the plane I am on while writing this, going from Dallas to El Paso—the rise is, at least at times, quite exhilarating.

Projecting the GIGA to the Year 2050

Biological populations reach limits imposed by their supporting resources, or are kept in check by predators. Algae are boxed by the edge of the pond, and then winter comes to kill them. Bacteria bump into the edge of the Petri dish and then run out of nutrients. Rabbits in Australia grow into a pestilence that triggers poisoning campaigns. Similarly, there will be limits to Earth's capacity to field the GIGA, and many environmentalists and ecological economists have called out warnings about such limits. Almost certainly oil is going to hit a wall of production capacity within a decade or so.

But human ingenuity can innovate. Human technological systems can create mutations that can overcome limits that previous generations of industry would have faced. The persistence of the GIGA so far leads me to think it will continue for some good amount of time, and the desire of people for more—indeed the dire need they have for more in many impoverished sections of the world—means that people will continue to participate in wealth-generating cooperatives, from small-scale entrepreneurial activities to globe-spanning multinational corporations. So let's see what happens when the GIGA is projected into the future. The year 2050 seems like a reasonable point, because it is within range of being experienced by most people alive today (even if not by me—I would be 100 in that year, but maybe I'll be fortunate).

Projections can be hazardous to the professional health of an analyst. Nevertheless, I have already made this extrapolation in figure 7.5. The GIGA, if it continues at 3.1 percent annual growth rate, will reach in 2050 a point at which the gross world product is about $240 trillion (in 2005 dollars). For comparison, in 2005 the GWP was about $58 trillion in that year's dollars. The increase is a factor of 4.

Now, the number I think is crucial is the GWP per capita. That number gives us a metric of the economic well-being of the average person. To elaborate on a general point I alluded to earlier, Vaclav Smil and other energy analysts have emphasized that we should really look at such indicators as low infant mortality, life expectancy, and education as pointers to economic well-being, and not a crude lumped metric such as GDP, which includes luxury cars and $100 bottles of liquor.[6] There is also a movement for alternative indicators, such as the gross happiness index being promulgated in the small Himalayan country of Bhutan. I am not here to defend the GDP or GWP as an indicator, and my purpose is not to examine what makes an adequate level of well-being. The fact remains that today politicians in the developed and developing countries get elected (in democracies) by promising that their policies will increase economic well-being, and people respond and clamor for more things, thinking those things fulfill needs and desires. In fact it is often the rich who are most desirous of even more money. We can project that people overall will support trends that make the GDP of their country and thus the gross world product grow. In 2005 the gross world product per capita was about US$9,000. What does it become when the GIGA is projected to 2050?[7]

To compute the projected GWP per capita, we need the population for 2050. Right now the official "best" projection by the United Nations is about 9 billion. This future number has been coming down, because birth rates have been declining to rates that a few decades earlier would not have been predicted. I personally think we might "merely" reach 8 billion by 2050, optimistically nudging the official equation down. With these two possibilities (admittedly not that different) for the global population, our projected GWP per capita in 2050 would be somewhere between $27,000 and $30,000 (again, in 2005 dollars), about 3 times what it was in 2005.

That is an interesting range of numbers for the year 2050. One might say: "That's crazy! Will never happen! A GWP that is 4 times today's? And a per capita GWP that is 3 times today's? Think of all the extras that amount of growth will entail: the coal mining, the strip commercial malls, the roads that need servicing, the electrical lines and cell phone towers, the factories, the houses. There will be mass extinctions of creatures! Pollution will be everywhere, and noise and more noise!" Those things may be true, but here is the point on which I wish to dwell: This projected GIGA leads to a world in 2050 in which its average citizen has a gross domestic product that is approximately equivalent to that of the average person in Japan or in the European Union *today*. And that average in 2050 would still be *below* the average in the United States today! Remember, we are talking about 2050, more than four decades away. And "average" doesn't mean everyone in the world. If the economies of the United States, the European Union, and Japan keep growing, they will remain above the average, which means that many places will be below the average. Even in 2050, the average citizen in many places will not have the economic well-being of the average European or Japanese today.

Thus, I conclude, a continuation of the present trend in the GIGA until at least 2050 is desirable because this growth rate seems a minimal rate to get the world's people up to a decent standard of living by 2050.

To throw support behind a slower growth rate, from the viewpoint of, say a reader in the United States, means to approve a vision of the future in which billions of people remain in poverty. Indeed, one might morally wish that the trend to reach the GWP's projected value in 2050 could occur even more quickly, because in places such as Rwanda and Bangladesh, where people are scrambling for firewood to cook with, much of the world is in desperate poverty. People are hungry. There are few opportunities for jobs. People suffer from lack of health care, and they are not in possession of what are considered basics of life, such as enough food to eat, decent housing, clothing and shoes, clean water, household lighting, clean cooking fuel, and education for their children — not to mention telecommunications, electronics, motorized personal transportation, and exotic vacations (which I, for one, enjoy). My projection, which I have made as simple as I can, is somewhat more optimistic than but roughly along the lines of the "business as usual" scenario of the Intergovernmental Panel on Climate Change's Third Assessment of 2001, which projected 2.9 percent growth until 2025 and then 2.2 percent growth until 2100.

By the way, the current average per capita GWP can be visualized approximately (within 15 percent) as the conditions in countries such as Brazil, Bulgaria, Iran, Mexico, Romania, and Turkey. (China had reached only 76 percent of the world average in 2006.) So in the scenario I am projecting and arguing as desirable, the world average level of wealth changes between 2005 and 2050 from a situation like that in present-day Mexico or Turkey into one like that in present-day Europe and Japan.

The Future of Energy and CO_2 Emissions

If the gross world product goes up by a factor of 4 by 2050, and because GWP, energy consumption, and CO_2 emissions are so correlated, we might predict that in 2050 the global values for both energy consumption and CO_2 emissions will also go up by a factor of 4. Fortunately, that almost certainly will not be the case. We can gain insight into why this is so by looking at past numbers for the three parameters.

Figure 7.6 shows the normalized numbers for GWP, energy, and CO_2 emissions for 1970, 1985, and 2000.[8] I have normalized all three to be equal to the number 1 in 1970, for reference, and thus the bars for 1985 and 2000 show the changes relative to their 1970 values. We can then easily see how the values are changing proportionately to one another as they grow.

It is evident that GWP rose fastest, followed by energy consumption, followed by CO_2 emissions. The lesson we learned earlier about how the bars could be at different levels for the EU and Japan relative

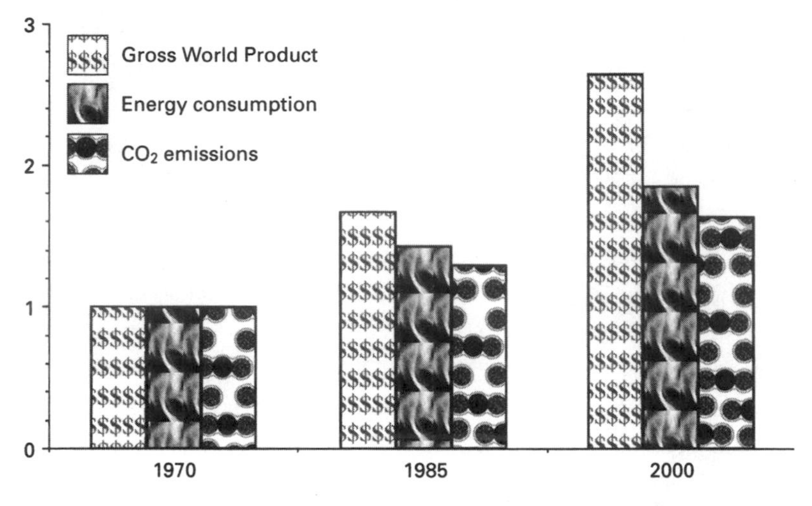

Figure 7.6 Gross world product, energy consumption, and CO_2 emissions, all as global values (not per capita), relative to their values in 1970.

to those for the United States is also true of the world over time. It looks as if the world has become more efficient in using energy to generate economic wealth over time, and more effective in the ability to create energy by generating fewer CO_2 emissions.

CO_2 emissions rose a bit more slowly than the increases in energy consumption because nuclear power or hydropower developed during this time do not emit CO_2 and also (the primary reason) because of the increasing use of natural gas as a fossil-fuel source. When natural gas (methane) is burned, the four hydrogen atoms for every carbon atom (figure 4.3) serve as a significant source of energy. Those oxidized hydrogen atoms become water (H_2O), and the process yields a significant addition to the energy that comes from oxidizing the methane's lone carbon into CO_2. On the other hand, coal (figure 4.2) has fewer hydrogen atoms than carbon atoms. In other words, the ratio of hydrogen to carbon is lower in coal than in natural gas. For a given yield of energy, coal will release the most CO_2, oil the next most, and natural gas the least. Therefore, the fossil-fuel "mix" for a given amount of energy affects the CO_2 emissions. As can be seen in figure 4.4, that mix changed as natural gas went from negligible amounts in the fossil-fuel mix in 1970 to about 20 percent of emissions in 2000.

Will this trend in the decline of "carbon intensity" (the ratio of CO_2 emissions to energy) continue? As the energy analyst Marty Hoffert and others have been emphasizing, the trend could soon start to reverse as coal usage is growing fast, particularly in China, and even more so in the future if coal starts being developed as a source of liquid fuels in places such as the United States when global oil production plateaus and then starts to decline. On the other hand, if nuclear power were to undergo a resurgence, or if the relatively rapid growth of wind and solar energy continues enough to transform those kinds of non-CO_2-emitting energy into truly global players from their current "bit parts," the decline in the carbon intensity trend could hold.

However, the change in the carbon intensity has been the smaller factor in the reasons why the rise in the recent historical emissions of CO_2 has been not as large as the rise in gross world product. The main component, as evident in figure 7.6, has been the change in the ratio of energy to GWP.

Energy costs money and is used for virtually everything. (Amazingly, in the United States gasoline retails for less than bottled water in a delicatessen in New York City, but that may say more about the commoditization of water than about the price of energy.) Manufacturers are driven by competition to reduce costs, and energy is one of the costs often targeted for improvements. Send in the engineers, please! Energy efficiency can be accomplished by making the conversion processes more efficient (e.g., by recycling waste heat in power plants) or by reducing energy losses (e.g., by improving the designs of refrigerators). So one part of the smaller rise in energy relative to GWP (or GDP in the case of individual countries in which the action of improvements actually takes place) is the generation of useful transformative power with less primary energy, say in the making of stainless steel spoons or the smelting of aluminum ore. A second part in the slower rise of energy relative to GWP has been the gradual shift in the mix of the world's economic sectors that contribute to GWP. I refer to the rise in service industries in comparison to, say, heavy manufacturing. The best information seems to indicate that the contribution from this "sector" shift has been about equal to the contribution from the "efficiency and conservation" shift. As nations become more developed, services grow, and services tend to require less energy per unit of economic dollar value. It isn't merely a matter of a developed country transferring its heavy industry to foreign manufacturers, because that does not necessarily result in a global net reduction in energy relative to a unit of GWP. What we have here is a global trend in the overall rise in services—not everywhere equally, of course, but on the scale of the global sum.

Can the historical rate of improvement (decrease) in the global ratio of energy to GWP continue? Could the rate of improvement get even better? This is a crucial question for Earth's future. Here I mostly want to highlight its significance and the fact that the answer is hotly debated by technical experts.

There will be thermodynamic limits to the energy it takes to drive certain kinds of transformations (in the most general sense). One such limit is the Carnot limit for the production of motive power (say, to turn the turbine) in a power plant that uses heat as the source of the power. But, as is clear in the analysis of the huge amount of waste heat relative to useful light in an ordinary incandescent light bulb, many aspects of human life can be targeted for improved energy efficiency and conservation. The growing number of hybrid cars on the roads could be one of those technology shifts that will play into keeping the trend of the ratio of energy to GWP in a decline. Recently I was in Turkey. After being directed to the rest room in a quaint, out-of-the-way restaurant, I began walking down a dark hallway. Suddenly a light bulb went on in the hallway, triggered by a motion detector. I was impressed. So long as the energy it takes to manufacture the motion detector is less than the energy saved in having the light off when no one is actually walking down the hall, the little detector is one of the "atoms" that are already taking a role in improved generation of wealth with energy (in this case by saving energy).

Can we project continued improvement? We can certainly try. For projected emissions in the year 2050, here are the numbers. They are quite close to what the Intergovernmental Panel on Climate Change projects for the "business as usual" scenario. I am using a simplified analysis that merely extrapolates large-scale trends, given the rationales above for reasons behind the trends. These trends have been historically afoot. Here are the results if they continue: To achieve the gross world product of nearly $250 trillion in 2050 (I cited $240 trillion above), the world will require the consumption of energy at about 2.5 times the rate in year 2000. And the CO_2 emissions for that energy will be about twice the emissions in year 2000. That is what it will take, projecting current trends, to lift the average standard of living in the world to a state about equal to that enjoyed by citizens of Japan and Europe today. It is cheery news that to nearly quadruple the gross world product, CO_2 emissions will only have to double. But then again, I repeat, the emissions are projected to double.

How high will atmospheric CO_2 go? To discuss that, I must return to the dynamics of the global carbon cycle.

8

How High Will the CO_2 Go?

If you are the kind of person who looks forward to each year's being special, the CO_2 data give you that—each year, another world record! Furthermore, the data are in, and they are unequivocal. The rise is caused primarily by the gas wastes from the myriad ways we combust the triumvirate of fossil fuels, augmented by the flux of CO_2 from land-use changes. And as the gross world product heads toward nearly quadrupling in 2050 relative to the century's early years, the world's people will need more energy. Fossil fuels are currently the energy source of choice, which means more CO_2 emissions. So the CO_2 will continue to rise. Can we predict by how much?

The Carbon Atoms Travel in the Pools of the Biosphere

With some events not described earlier in the book added to those already detailed, here is a summary of Dave's story:

32,000 years ago: entered the biosphere from limestone dissolution, then statically typical adventures in the biosphere for 35,000 years

1964: passed through infrared gas analyzer at Mauna Loa Observatory

1967: in ocean as bicarbonate
1971: in body of marine diatom
1971: in body of zooplankton that ate the diatom, then respired back into seawater; became bicarbonate
1979: in atmosphere
1981: in ocean as bicarbonate
1983: in atmosphere
1984: in ocean as bicarbonate
1985: in body of plankton
1987: respired by marine bacterium, so in ocean bicarbonate again
1993: in atmosphere
2003: in rice grain in Bangladesh, in same plant that contained Coalleen in its stalk
2004: released into atmosphere by exhalation within a day after post-dinner digestion of cooked rice.

The global cycle of carbon is not simply like the turn of a Ferris wheel or a commuter's round trip to work and back. Instead, the pathways of carbon are more like strands of cooked spaghetti flung toward several deep serving bowls on a table. The strands go in and out of various bowls, and they swirl around within the bowls. Many of their paths merge into loops. A strand goes from one bowl to another, and from there a second strand goes to still another bowl, which might be, in fact, the original bowl or a third bowl from which you could hop a strand to go back to the first bowl or glide along yet more strands to go to a number of possible fourth bowls or then fifth or sixth ones and from every one of those go back to any of the others and eventually even back to the first.

Here is a summary of Coalleen's story:

1964: entered the atmosphere (and thus the biosphere) from coal combustion
1965: in fig tree in South America
1967: dropped as leaf into soil litter

1970s: entered soil layer beneath litter
1979: digested by organisms, excreted into soil, then diffused back into atmosphere
1985: in oak tree leaf but then respired within a week back into the air
1987: in ocean as bicarbonate
1991: moved by ocean currents into intermediate water just beneath the surface mixed layer
2001: moved by currents back to ocean's surface, then popped into the atmosphere
2003: in rice stalk in Bangladesh, in same plant that contained Dave in a rice grain
2004: released from stalk when burned to cook dinner in Bangladesh.

How are the bowls defined? The greatest bowls—atmosphere, soil, ocean, and life—are physically distinct. Furthermore, the strands that swirl around inside these great bowls pass in and out of smaller internal bowls, creating a nested hierarchy. The zones within nature provide many of the definitions for the nesting. The fresh litter that blankets a forest floor is a distinct smaller bowl of detrital carbon, a layer within the larger bowl of "the soil." The first 100 meters of depth (more or less) of the ocean constitute a distinct "mixed layer" in which the water's chemical properties are homogenized by the wind, forming a well-defined smaller bowl within the 40-times-larger volume of the whole ocean. The bowl of "all life" can be partitioned into terrestrial life and marine life. The next level down in the nested sequence within those two bowls contain the often squirmy, usually sexy, and quite pretty smaller bowls of individual creatures, sometimes lumped together by their ecological functions of photosynthesis, consumption, and decomposition. Thus, the smaller bowls within the bowl of marine life are the phytoplankton, zooplankton, and bacteria, a division that leaves out the fish and the whales (no insult to them intended, and they could be folded in with the zooplankton as consumers) but which does cover most of the carbon fluxes.

Oiliver's story is as follows:

1964: put into the air (and thus the biosphere) from gasoline combustion
1966: in ocean as bicarbonate
1970: in atmosphere
1976: in prairie grass in Australia
1976: in body of kangaroo
1983: exhaled into air by kangaroo
1989: in ocean as bicarbonate
1992: in atmosphere
1994: in acacia tree in Rwanda, in cellulose next to Methaniel
2004: released into atmosphere from acacia tree when burned for cooking in Rwanda.

Collectively, the great bowls of atmosphere, soil, oceans, and life constitute the biosphere. I am tempted to call the biosphere the table that holds the bowls, except for the fact that the four great bowls *are* the biosphere, which is not a table but rather the greatest bowl of all and which connects to the subterranean bowl of the rocky planet underneath the thin film of this active surface biosphere. Most of the strands I have portrayed so far as ins and outs of each of the biosphere's four great bowls come from and go to sub-bowls or sub-sub-bowls within the great bowls of the biosphere. Only a few strands, infrequently traveled by the atoms, enter or leave the encompassing bowl of the biosphere itself, such as when Dave and Icille emerged from limestone 32,000 years ago and when Coalleen, Oiliver, and Methaniel were "liberated" from fossil fuels, participating in a flux that was detailed in chapter 4 to be already 20 times the natural flux of "fresh" carbon atoms into the biosphere.

Here is Methaniel's story:

1964: entered the atmosphere (and thus the biosphere) from the combustion of natural gas
1969: in tundra plant, then respired by the plant right back into the air
1972: in ocean bicarbonate
1976: in atmosphere
1978: in leaf of cypress tree in Louisiana
1978: in soil litter
1981: respired by soil bacterium, back in air
1987: in ocean as bicarbonate
1988: in phytoplankton, then in zooplankton, then in small cod
1990: in shark body, then exhaled and in ocean bicarbonate
1992: in atmosphere
1994: in acacia tree in Rwanda, in cellulose next to Methaniel
2004: released into atmosphere from acacia tree when burned for cooking in Rwanda.

As I illustrated in the preceding chapter with a few specifics and mostly with statistics, during Dave's travels over the 32,000 years since he emerged from limestone, he spent enough time in the biosphere to have visited all the major bowls and most of the sub-bowls a number of times, and he experienced virtually all the chemical forms of different molecules that a carbon atom can "enjoy" when in the biosphere, including lots of time in the deep ocean. So far, in the last 40+ years none of our carbon atoms have gone down into the ocean's abyss, but eventually they will. And when they do, they will be gone a long time (1,000 years on average), but even there they will still be within the biosphere and thus destined in that geological finger snap of a millennium or so to come back up and enter the more active parts of the biosphere again, in atmosphere, in surface ocean with marine creatures, and in land ecosystems.

Here is a summary of Icille's story:

32,000 years ago: entered the biosphere from limestone dissolution along with Dave; after a few quick adventures (e.g., being inside a lion), was trapped in a bubble of air in an Antarctic ice dome

1994: liberated from ice bubble when a slice of ice was melted and analyzed; released into the atmosphere.

In short, the fossil-fuel atoms do not stay in the atmosphere into which they were ejected by industrial fossil-fuel energy servants. Neither do naturally released atoms, such as Dave and Icille, stay in the air, or in any active pool. They *all* circulate. They *all* contribute to the chemical activities to which carbon atoms belong in whatever pool they are in. To paraphrase Gertrude Stein, a carbon atom is a carbon atom is a carbon atom. We must understand these dynamics of carbon before we can make a good projection of future CO_2 levels.

The Airborne Fraction over 40 Years

As was detailed in figure 5.1, in the period 1966–2005 the atmosphere's CO_2 grew by 130 billion tons of carbon in the form of CO_2. During that same time, injections up into the air from fossil-fuel combustion, including a small component from other industrial processes such as cement production, totaled 225 billion tons of carbon in the form of CO_2. Therefore, fossil-fuel combustion alone can account for the rise in the air's CO_2, if 58 percent of the fossil-fuel CO_2 stayed up in the air during those 40 years.

But another source was noted. This source, also hinted at in the discussion of whether cutting trees for cooking fires would be sustainable in countries such as Rwanda, was "land-use change," including fluxes to the air from deforestation and conversion of land to agriculture as well as return fluxes from the air during reforestation. For the 40 years cited above, land-use change was a net positive flux, and one number for the total flux during this time period in the scholarship of the global carbon cycle is 75 billion tons of carbon in the form of CO_2, which is a nontrivial increment to the amount from fossil-fuel combustion. It might be smaller, and there is some controversy about this amalgam number (which has to take into account a wide variety of changes across many places all around the planet), but for now combining the fossil-fuel flux with that from land-use change gives a summed flux over the 40 years of 300 billion tons of carbon, of which the land-use flux is about 25 percent.

Thus, of the 300 billion tons of total flux, 130 billion tons of carbon "remained" as excess in the atmosphere as the air grew by slightly more than 60 ppm over the 40 years. The ratio of these two numbers (130/300), called the *airborne fraction*, is the amount by which the atmosphere actually grew in CO_2 relative to the emissions it received. For the 40 years cited, then, the airborne fraction was 43 percent. The other 57 percent of the emissions went somewhere else in the biosphere other than the atmosphere, even though all the emissions started out by going up into the atmosphere. The airborne fraction will be used to make a projection about the future rise of CO_2, but right now it should be regarded as only a lump-sum number across 40 years of time.

Where does the CO_2 added to the air go? How constant are the absorption processes, whatever they are? What will the airborne fraction be in the future—say, in 2050?

Equal Exchange Does Not Create Net Change

The flux of CO_2 from fossil-fuel combustion is a true asymmetric flux: it goes from fossil-fuel energy servants into the air. The flux coming from changes in land use, however, is a net flux that results from unequal amounts of two fluxes going in both directions: roughly, the flux from deforestation and agriculture that goes into the air, relative to the flux from the air back to land ecosystems during reforestation and agricultural soil restoration.

There are one-way and two-way fluxes in the natural carbon cycle as well. Carbon flows from plants into herbivores, for example, and not at all the other way (except by the loops through other pools, of course). But in the large-scale system that concerns us here—the two largest bowls in contact with the atmosphere, namely the oceans taken as a one whole and the land ecosystems taken as another whole—the fluxes go both ways back and forth with the atmosphere. To get a sense of what happens to the carbon once it enters the atmosphere from the anthropogenic sources, it is necessary to inquire into these bi-directional exchanges in the global carbon cycle.

There is a crucial point here, as I noted in passing in chapter 5: bi-directional exchange alone can dilute the fossil-fuel CO_2 in the atmosphere, but it does not cause a net removal of that excess CO_2 that is today nearing 40 percent above its pre-industrial amount.

First, consider dilution. Because exchanges go back and forth between bowls, the atmosphere's fossil-fuel carbon will get diluted into the other bowls and will be replaced, to some extent, with carbon from the other bowls. Coalleen, Oiliver, and Methaniel all left the atmosphere within a few years of their first release into it in 1964. But at various times later in their travels, they all reentered the atmosphere. In fact, they left and reentered several times. So what is going on when we try and figure out the numbers that say that 43 percent of emissions stayed in the atmosphere and 57 percent went somewhere else in the biosphere?

An analogy might help. Brew a full cup of fresh, dark tea. Now place the cup in a sink below an open faucet's trickle. Already full at the start, the cup that now runs over undergoes an exchange in which the trickle of clear water that comes in from the faucet equals the overflow of tea-colored water that leaves the cup. Slowly the liquid loses its rich tea color. The tea becomes diluted. Yet during all this time the cup remains full. Its volume has not changed. Mere exchange alone is not a process that alters the volume of fluid within the cup. Similarly, mere exchange alone does not alter the amount of carbon within any bowl of the biosphere.

I do not mean to claim that the atmosphere and the other carbon pools are like cups of overflowing tea, because they are not full and they do not overflow. The tea is only an example of how exchange creates dilution but not net flux (in other words, removal of the excess CO_2). For net flux that changes the level, the exchange must be asymmetric

In the atmosphere, for instance, if fossil-fuel CO_2 molecules are exchanged one for one with CO_2 from the ocean, back and forth, 100 billion tons of carbon per year up and 100 billion tons per year down, the concentration of fossil-fuel atoms in the atmosphere quickly falls as more and more of them go into the ocean. But if they are replaced with CO_2 molecules from ocean, the level of excess CO_2 in the air, which began as input of fossil-fuel CO_2, remains the same, even though that excess is not itself made of the original fossil-fuel molecules. This is a bit tricky, but the tea analogy is a good way to visualize the process. The same goes for the air's exchange of carbon with the land's organisms. If plants remove from the atmosphere 60 billion tons of carbon as CO_2 per year and the soil microbes and other terrestrial respirers breathe out 60 billion tons of carbon per year that go back into the atmosphere, then the net removal of CO_2 is zero; we are left with the same excess of

CO_2 in the atmosphere, only with some of the excess replaced by CO_2 derived from respiration.

Consequently, in the real world, because 57 percent of a mass that was equivalent to the added flux of fossil-fuel CO_2 (I am being careful with the wording here) disappeared somewhere during the 40 years of this calculation, more than mere equal exchange took place between the air and the other bowls. There must have been a net flux flowing from the atmosphere to at least one of the other great bowls of the biosphere. The elevated CO_2 in the atmosphere must somehow be forcing an extra flux into either the ocean pool or into the combined pool of land plants and soil, or into both pools, above the mere exchange of equal fluxes that would be the case during times when the natural carbon cycle was in balance (for example, from 12,000 years ago until about 1850).

Carbon Is Pushed, Carbon Is Pulled

What makes the various molecules travel along pathways between bowls? There are two basic options. The molecules are either pushed out from a bowl or pulled into a bowl. For example, the burning of fossil fuels pushes CO_2 out into the atmosphere. During photosynthesis (for instance, when the barley plant grabbed Dave from the air), CO_2 is pulled from the atmosphere.

Pushes often come from the natural jostling of molecules. The more jostling (which is usually a function of concentration or temperature), the greater the oomph behind the push. When gas molecules are pushed from the air into the ocean, how many gas molecules make the crossing in a given time period depends on the concentration of the specific gas. The same goes for the rate at which dissolved gases are pushed back from the ocean into the air.

Right now, the atmosphere is being artificially pumped up in its CO_2 pressure by the ongoing fossil-fuel input. Because the rate at which CO_2 is pushed from air to ocean depends on the concentration of CO_2 in the air, we would immediately expect the following result: Overall, the ocean is a net sink that captures some of the excess CO_2. As the atmosphere becomes carbonated, so to speak, more flows into the ocean than comes up out of the ocean. A net amount—not just dilution of fossil-fuel CO_2, but a true net amount of mass—must, by the laws of the physics of gases, leave the air and go into seawater.

Another form of pushing comes from the active exhalation or excretion of CO_2 as a waste gas from any creature that respires. When soil bacteria excrete their CO_2 by-product outward into the interstices of the soil, the pressure of the gas builds up in the soil. That, in turn, creates a pressure due to jostling that drives the CO_2 up into the air.

On the pull side of things, the most significant pulls come from organisms feeding on various forms of carbon. When you spoon ice cream into your mouth, you are pulling carbon from the ice cream bowl (sometimes literally) into the bowl of the human body. Because bacteria must feed, the soil bacteria, which as a collective bowl excrete waste CO_2, pull organic carbon into their bodies from the detrital bowl of the soil. A hawk snags a mouse, pulling carbon from the bowl of prey into the bowl of predators. And, as noted, photosynthesizers pull carbon either from the bowl of air (in the case of plants) or from the ocean's bowl of carbon ions (in the case of phytoplankton), thereby gaining the most important building block for their bodies.

Analyses of the uptake of CO_2 by the oceans cannot account for all of the net flux out of the atmosphere. Carbon-cycle scientists also finger land plants as players in an asymmetric exchange with the air. What drives this? Because all plants require CO_2 as a body-building element, they might grow better, and thus pull more from the air, when there is more CO_2 in the air. This "CO_2 fertilization" has been confirmed by many experiments, the most elaborate of which go by the acronym FACE (standing for Free Air Carbon Enrichment).[1] The main finding from these, as well as from experiments in greenhouses and growth chambers, is that most plants do better under higher CO_2 levels. But there are large differences among species. Differences also appear along stages of the life cycles of the plant species. Both phenomena cause ecologists to worry that under ever higher levels of CO_2 the mixture of trees and herbs in natural ecosystems will change, favoring responders to elevated CO_2 and disfavoring those species that benefit only slightly. There is also evidence that the protein content will drop as plants require fewer enzymes in their leaves to help with photosynthesis, which could affect the diet and feeding habits of animals that feed on plants.

We now have reasons why there might well be net flows of CO_2 from the atmosphere to the ocean, and from the atmosphere to the plants. That such net fluxes exist has been confirmed by experimental measurements and by mathematical models that interpret the measurements, although many details are still being worked out and analytically debated in the technical literature.

The Airborne Fraction: Jagged from Year to Year, More Constant across Decades

Predicting the future of the air's CO_2 requires a prediction about the future of emissions (see chapter 7) and then a prediction about the future of net exchanges with the ocean and land ecosystems. To ground the second prediction, it helps to scrutinize the past in more detail.

I have plotted four lines in figure 8.1. One shows the annual fossil-fuel emissions. Another, the uppermost, sums the total emissions (fossil fuel plus land use). Yet another tracks the annual measured increase

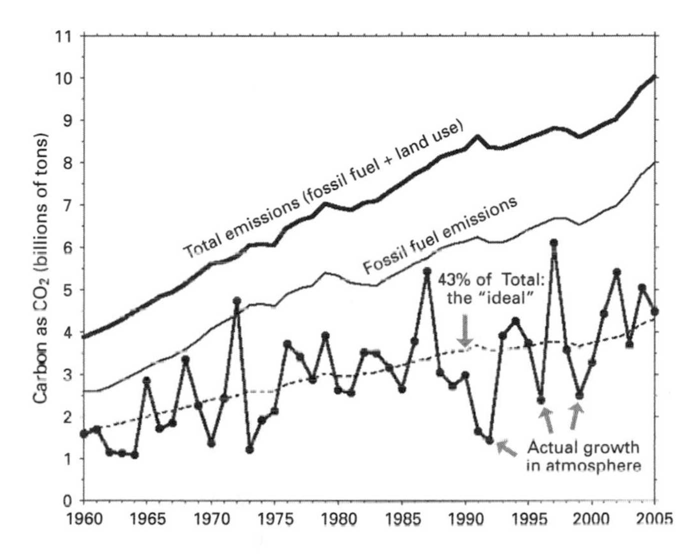

Figure 8.1 The annual actual growth rate of atmospheric CO_2 varies substantially and is clearly different from a 40-year average of 43 percent of total emissions (the "ideal").

in atmospheric CO_2. The broken line, which I call the "ideal," does not represent a measured quantity, as the three other lines do. The "ideal" shows what the air's increase would have been if during each year the dynamics of the carbon cycle behaved like the overall average over the time period—in other words, if each year's increase in the air's CO_2 had been 43 percent of that year's total emissions. This allows us to compare the actual yearly increase with an ideal yearly increase.

It is clear that on a year-to-year basis the dynamics of the air's growth do not conform to the long-term average of 43 percent of total emissions. Indeed, in extreme years the growth rate can be either nearly double or just half the long-term average. In fact, the growth rate seems quite erratic from year to year.

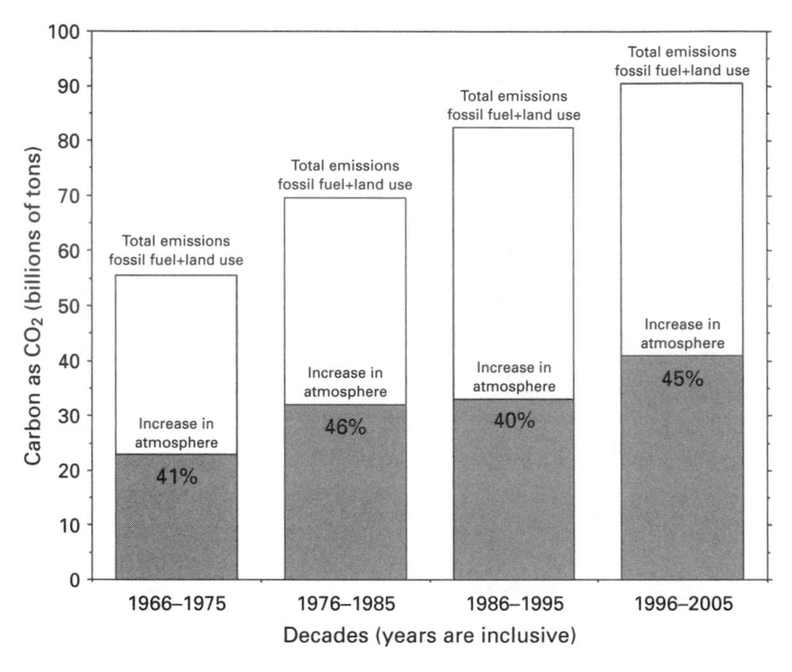

Figure 8.2 The shaded portion of each bar is the ten-year increase in atmospheric CO_2, with the percent giving the fraction of the increase relative to the total emissions.

What happens if the analysis is performed across longer time spans? In figure 8.2, to address this question, I present the above data in ten-year chunks. To simplify, I show just the total emissions as bars for the given decades, and I give the observed airborne fraction for each decade as a numerical percentage in the lower shaded portion of each bar, which shows the actual atmospheric increase during each decade.

Ah! Now the biosphere system appears much better-behaved. Ten-year chunks of the air's increase deviate only by a few percent from the historical average of 43 percent of total emissions. This is cause for hope that the future of atmospheric CO_2 can be predicted, given a projection for the emissions themselves.

The Constant Airborne Fraction

I now want to see if a stripped-down "calculation for dummies" can replicate the growth of atmospheric CO_2 seen in the records, say at Mauna Loa or Antarctica. (For present purposes we can assume that the annual averages are the same, so which data set we choose doesn't matter at the level of detail here.) I am also going to cheat, in a sense, by using the record itself to pin both the start point and the end point of the model. Here is the crucial step: I am going to assume that for each year a certain percentage of that year's emissions "stayed" in the atmosphere. The individual atoms of the emissions do not stay in the air long at all (as is evident from the convoluted trails of our named atoms), but the yearly rise can be conceptualized as a certain fraction of what was injected.

So here is the plan for the "dummies" calculation: Start with 1960. Assume a fraction to be applied each year of the total emissions. Add that to the model atmosphere's CO_2 level. Do the same for the next year's emissions, and the next, all the way. Does this model capture the path of the rise? If so, how well? If it does and does well, then perhaps the model can be used for predicting the future, assuming certain emissions.

This calculation is called a "constant airborne fraction model" because it assumes that each year a certain airborne fraction of the fossil-fuel emission is the amount that the atmospheric CO_2 increases. The results are illustrated in figure 8.3.

The match is impressive. The calculations from the model align well with the data. In fact the fit is remarkable.

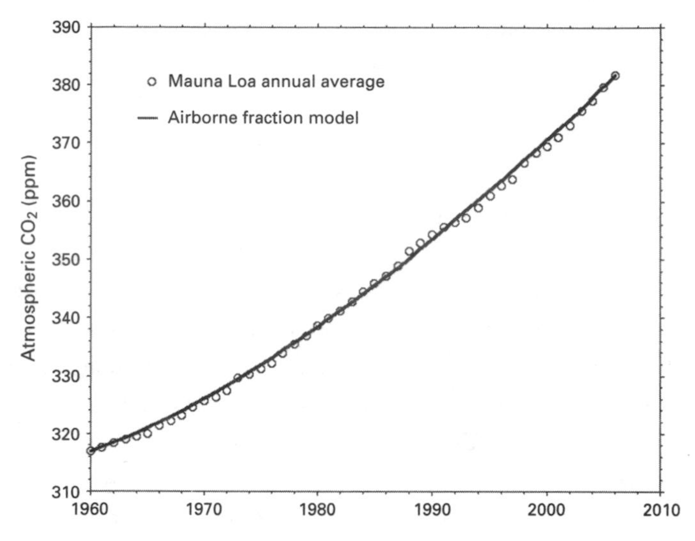

Figure 8.3 Data for the growth of atmospheric CO_2 (here, using the historical record from Mauna Loa as the data) are reproduced quite well by a model that uses the simple assumption that the annual growth is 43 percent of each year's total emissions into the atmosphere.

This simple calculation, the simplest possible "model" for the increase in the air's CO_2, matches the actual history of the rise almost exactly. Of course the end points had to match, because I took the fraction to apply to each year from the start and stop numbers that bound the time period. But that didn't imply that the entire time courses of the data and the model would match so well. If the airborne fraction changed significantly over time, we would see it, because in some long stretches the model would overshoot the growth rate of data and, compensating to make the end point the same, which I ensured, in other long stretches the model would undershoot the actual growth rate. But instead, the model and the data match closely.

The impressive fit demonstrates that the pools of the biosphere that exchange CO_2 with the atmosphere are behaving, together, like an absorption machine that takes up a certain fraction of the incremental rise caused by the release each year. Sure, that isn't true from year to year, as we have seen. But over longer time intervals the absorption machine is fairly well balanced. Times of low absorption are soon balanced by times of higher absorption. The biosphere behaves predictably, so far. We see no huge surprises from nature. We know there have been surprises in the distant past, as when changes in the natural cycles took CO_2 from the ocean and put it into the atmosphere at the end of the Ice Age. The reasons for this event are not conclusively understood, and there could be more surprises awaiting us in the future. But for now, we know at about what rate the atmosphere's CO_2 will go up in the future, if we can project the emissions themselves into the future (as was done at the end of chapter 7).

A word of warning: Care must be taken in using the constant airborne fraction as a way to extend a prediction. What we have here is not the carbon-cycle equivalent of gravity. Nor is the airborne fraction like Einstein's $E = mc^2$. We could posit that the net annual atmospheric CO_2 increase is equal to 43 percent of the total annual emissions, but (as can be seen in figure 8.1) that is most definitely false on an annual basis. A thought experiment may also be useful. Pretend that, in a fool's environmental heaven, next year's global emissions were zero. Then a calculation with any constant airborne fraction would predict a zero increase in the atmosphere's CO_2 that year. But in the "real-world" dynamics of this hypothesized state of zero emissions, the CO_2 would begin to drop. That's because right now the system is set so that the atmosphere possesses an excess and that excess is driving asymmetric exchanges with the ocean and land ecosystems. Those asymmetric exchanges would not disappear in a year.

Why even discuss the constant airborne fraction? Well, it does seem to hold over longer time periods. The airborne fractions were quite consistent over decades, though the reasons for the differences are still being sorted out by carbon-cycle scientists. The long-term record, computed year by year, allowed the ups and downs of the deviations of the annual airborne fractions from the "ideal" to balance out and replicate the data to an impressive degree. A recent paper suggests that the airborne fraction has been increasing slightly, but this result is statistically very weak.[2] Thus, the calculations shown here do indicate that, so long as the emissions are relatively high and growing, the assumption of a constant airborne fraction could give us a good prediction for the future of atmospheric CO_2.

In sum, the system is complicated but is behaving quite dependably. What causes the variability in the airborne fraction from year to year? That has been shown to be related to temperature and rainfall, for climate itself has fluctuations. A particularly strong correlation is with the cycles of El Niño and the changes thus produced in the productivity of tropical vegetation.[3] But on average, there is a certain absorption rate in the biosphere. Yes, the constancy is a bit mysterious, because the carbon cycle is so complex. Then again, the absorption by the ocean has a lot to do with physics, and even the plants on land are biophysical matter transformers that behave at least somewhat predictably.

I called this calculation that uses an airborne fraction a model. I did so with a wink. It doesn't really capture the complexities of the biosphere's carbon cycle. It is not a *dynamical* model, which would include actual equations that predict and track the flows of carbon along the strands between all the bowls: equations for gas exchange with the ocean, equations for photosynthetic uptake by land plants, equations for the release of CO_2 from bacterial respiration in the soil, and so on. I have made dynamical models, and others have made even more sophisticated ones, and they give about the same results as the "dummies" calculation with the constant airborne fraction. The point is that so far the bowls of the biosphere in contact with the air and with one another are acting as a fairly dependable CO_2 sponge to remove, on average, a constant amount of the total CO_2 source to the atmosphere.

Now I could go straight to a prediction about the future, by applying the constant airborne fraction to the probable course of emissions developed in the preceding chapter. But first let us see what more sophisticated models say about the dynamics of what is going on in the biosphere as the atmospheric CO_2 grows from the injected fossil-fuel CO_2 that subsequently disperses into the other bowls of the biosphere.

The Airborne Fraction from More Complex Models of the Global Carbon Cycle

In general terms, a scientific model is a mathematical description of a real-world system of processes. A model aims to imitate as closely as possible the interactions of those processes in the real-world system. Models are essential to developing and testing scientific understandings. A model can be as brief as a single equation, but often, especially these days, models are as complicated as all those strands of spaghetti going in and out the bowls of the global carbon cycle. Equations for the carbon cycle, for instance, aim to describe and then predict the major fluxes going in and out. A carbon-cycle scientist can then "run" the model and presumably watch how the carbon fluxes proceed from bowl to bowl, just as they do in the real world.

Most modelers would agree that the airborne fraction will increase somewhat in the future as the CO_2 rises. In other words, the absorbing powers of the other pools of the biosphere will decrease. As the surface of the ocean gains CO_2, its ability to absorb even more will depend on the replacement of that surface water with undersaturated deeper water. Furthermore, warming ocean waters will be able to absorb less CO_2. Enhanced plant growth from CO_2 fertilization is expected to be more substantial now than later, because plants' CO_2-uptake mechanisms gradually saturate at higher levels. It is like using mineral fertilizer on a farm. The first little bit does a lot, but adding the same increment doesn't accomplish as much, except establish an example of the law of diminishing returns. Furthermore, there are climate feedbacks upon the carbon fluxes of land plants and soils. In the early stages of warming, the extra temperatures might help by lengthening the growing seasons and by enhancing the overall physiological responses of plants (to different extents for different crops and in different climate zones.) But the warming also increases the metabolism of soil bacteria, which chew up the soil's organic store of carbon and release it as CO_2. It has been postulated that this could result in a large additional positive feedback flux of CO_2 to the atmosphere—especially from regions that have large amounts of soil carbon, for example in the form of peat.

So prediction ultimately requires models that follow all the strands and that follow how the flows along the strands will change over time, being influenced up or down by feedbacks from climate as the world warms. For comparison, eleven models all used the same emissions scenario for the next 100 years to project the rise in CO_2.[4] In the case in which the models ran without climate feedbacks upon the bowls of the carbon system, their average airborne fraction for the 100 years was 47 percent—not very different from the behavior over the last almost 50 years. But there were differences in the models. One standard deviation of the results (about two-thirds of the models) gave airborne fractions in the 42–52 percent range. That's fairly consistent. But the numbers also mean that some models were outliers from even this range. This shows the distribution of uncertainties, because top scientists constructed their own unique models using equations for the strands and the bowls as best they could.

Next, when climate feedbacks were allowed to change the dynamics of the carbon cycle in the eleven models, the airborne fraction went up, showing that as the world warms more CO_2 will almost certainly stay in the atmosphere. How much more? The average airborne fraction in the feedback case was 55 percent, with a standard deviation in the range of 46–63 percent. This is for 100 years, with the warming becoming more pronounced toward the end. The increase in the airborne fraction in the feedback case would be smaller if considered only to the year 2050.

The biggest differences among the models had to do with uncertainties in predictions about land plants and soils. That is not surprising, insofar as the response of the ocean is closer to pure physics. Roughly half of the CO_2 that leaves the atmosphere in these models went into the ocean, roughly half into the land ecosystems. But again, differences in some cases were substantial. And we would want to consider all those cases as possibilities.

There is no doubt that humanity is, in a real sense, gambling with the future. One part of this gamble has to do with uncertainties about the absorbing powers of the non-atmospheric parts of the biosphere to the CO_2 excess in the atmosphere. Acknowledging this, I will move on to a prediction, given the constancy of the airborne fraction so far over the last nearly 50 years and the fact that for the next 100 years the average behavior of the models, despite their acknowledged differences, was rather consistent with the behavior of the biosphere in the recent past.

CO_2 in 2050

Probably the most common rule of thumb for the absorption of CO_2 emissions by the ocean and terrestrial ecosystems is the figure of 50 percent. If the historical emissions from land use were those shown in figure 8.1, then the past airborne fraction was 43 percent. But the fraction is expected to rise in the future. So 50 percent seems reasonable, perhaps a bit overpredicted. However, if the land-use emissions were only half of the increment above the fossil-fuel emissions in figure 8.1, then the past airborne fraction was 50 percent, and we might argue that by 2050 it will not change that much, because the changes in the models as the feedbacks kick in are more pronounced in the second half of the current century.

The detailed dynamic models, too, confirm that 50 percent is about right. They averaged 47 percent over the next 100 years without the climate feedbacks on the carbon cycle, and 55 percent with the feedbacks, but, again, most of the feedbacks shift into gear in the second half-century. So we have evidence from the historical record and from the best current modeling that the airborne fraction in the future (say, 2050) will be about 50 percent. Of course this number has a plus and a minus, and the scientific carbon-cycle community will be watching what happens in the future and attempting to make the models more realistic.

Using a projected rate of input to the atmosphere that I will call the Central Trend-2050 projection of fossil-fuel emissions (see chapter 7), as the gross world product goes up about fourfold from what it was in the early years of the twenty-first century, CO_2 emissions approximately double. Past improvements in energy efficiency and conservation and structural changes in the economy generate wealth more efficiently

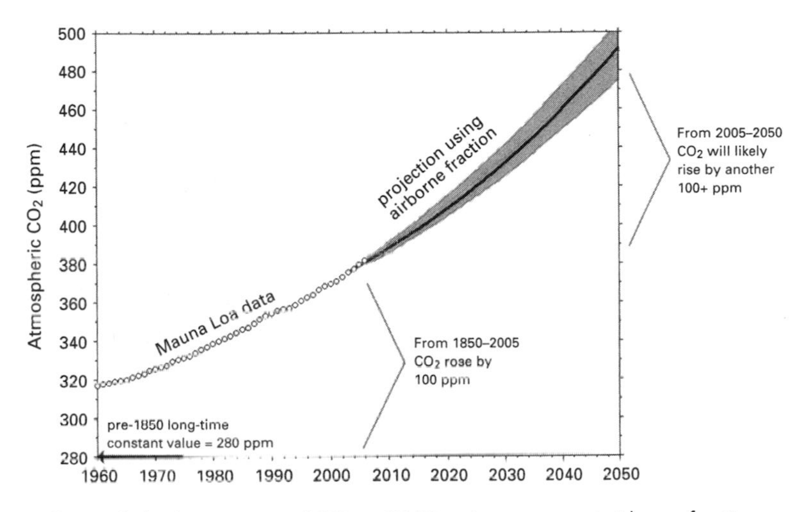

Figure 8.4 A projection of CO_2 to 2050, using a constant airborne fraction of 50 percent (45–55 percent for the band of gray), and the projection for the future of fossil-fuel emissions justified in chapter 7.

without requiring as much energy as in the past, and are assumed to continue as trends.

I am going to ignore land-use changes, because they are so difficult to predict. A number of different scenarios for emissions developed in the 2001 assessment of the IPCC considered that the land-use emissions, from 2010 to 2050, would be only about 5 percent of the total emissions.[5] That was primarily because fossil-fuel emissions continue to rise. The scenarios varied from zero to 10 percent in the fraction of total emissions that were from changes in land use.

Figure 8.4 shows the projection based on Central Trend-2050 projected emissions and the "nice" round airborne fraction of 50 percent, taking into account some degree of uncertainty with a band of gray (thus varying the airborne fraction between 45 percent and 55 percent).

The first thing that might strike the eye is that the projected rise in CO_2 from the early 21st century to 2050 will be about the same amount by which it has already risen above the pre-1850 value. Both the past historical increment and the future projected increment are about 100 ppm, which will put the CO_2 in 2050 about 200 ppm above its pre-1850 value. (Before 1850 it had been relatively steady for more than 12,000 years.)

Thus, the rate of increase is increasing—no surprise given that energy use will rise, though more slowly than the 3 percent exponential growth rate of the gross world product. Earth will continue to have an atmosphere with a CO_2 level higher than any shown by ice-core measurements stretching back more than half a million years. The game of geophysical gambling gets even more serious after 2050, but then we are stepping farther into the unknown with regard to the emissions. It seems reasonable that by 2050 the world's countries will have begun to get politically serious about coordinating their emissions in global agreements of some kind. Otherwise the biosphere will be shot more and more into unknown territory. Those of us alive for the next 40 years will likely see the concentration of CO_2 in the atmosphere increase by another 100 ppm.

9

Reining In the CO_2 Increase

The *climate sensitivity*, standardly used in predicting the course of global warming, puts a number on how much the global average temperature will change for a doubling of carbon dioxide from its pre-industrial value. The climate sensitivity can be used to roughly set the level of anxiety or complacency, given a sense of what a given temperature change means for rainfall, the seasons, soil moisture, extreme weather events, and all other variables that affect our lives and the lives of other creatures in Earth's various habitats.

Other greenhouse gases, including methane and nitrous oxide, are rising too. These should be folded into climate predictions as forcing factors that contribute to warming. The same goes for the several kinds of forcing factors that act as cooling agents—for example, sulfur aerosols that float in the sky and reflect sunlight, primarily from coal-fired power plants. The list goes on: carbon soot in the air that absorbs the incoming sun (different from CO_2, of course, which absorbs Earth's outgoing thermal infrared radiation), and increasing levels of low level ozone, a pollutant that also acts a greenhouse gas and is to be distinguished in its effects from the vulnerable stratospheric ozone that absorbs biologically damaging ultraviolet radiation from the sun. And others that are more minor. Climatologists attempt to understand the combined effects of all of these agents as a system of perturbations.

General scientific agreement has it that over the next 100 years the biggest forcing agent on climate—probably by 90 percent—will be CO_2. That is the reason for the intense focus on CO_2 in so much of the political and economic hubbub that is stirring with ever-increasing vigor—carbon offsets, carbon taxes, carbon cap and trade, carbon treaties. Before about 1850, CO_2 levels were stable at about 280 parts per million in the atmosphere for thousands of years. Thus, the value for doubling is 560 ppm, which, extending the Central Trend projection that concluded the preceding chapter, would be reached sometime around the year 2070.

The most skilled modeling efforts of climatologists have historically been able to peg the climate sensitivity only to within a factor of 3. This uncomfortably large degree of uncertainty is predominantly due to the trickiness in quantifying the effects of climate change on clouds, which both reflect sunlight and also act as infrared absorbers. Some climatologists hold that the uncertainty can be narrowed by the analysis of past ice-age climates, because ice cores and other evidence provide the CO_2 levels and the approximate global temperature. The most recent technical evaluation of the climate sensitivity comes from the 2007 reports of the Intergovernmental Panel on Climate Change.[1] The value is 3°C (5.4°F) for a doubling of CO_2. The uncertainty in this claimed by the IPCC is slightly more than a factor of 2, from a lower bound of 2°C to an upper bound of 4.5°C. (The range in °F is therefore 3.6–8.1.)

The number 3°C (5.4°F) may or may not sound like much to you. Some people would even welcome higher temperatures at certain times of the year. New York City had an amazingly warm December in 2006, and most environmentally oriented folks I knew saw the warmth as a harbinger of worrisome global warming but also reveled in it. (It lasted only a month before a frigid winter.)

To my mind, the best way to develop an intuition for what an average global shift of, say, 3°C will mean for the future is to consider the fact that in the coldest depths of the last ice age, about 20,000 years ago, Earth's globally averaged temperature was "only" about 5°C colder. Climate change is amplified in the high latitudes, and during the ice age thick ice sheets extended all across Canada and into the northern portions of the United States. A correlated expansion of continental ice occurred in Eurasia. Sea level was more than 100 meters lower because of the water locked up on land in those ice sheets. Furthermore, the vegetation of the Amazon Basin was not a continuous rain forest but much more patchy. Across the world, the distributions of species of trees and grasses, and the animals associated with different ecosystems, were not only shifted farther south but responded to the Ice Age climate by producing different ecological mixes of species than occurred in the post-glacial age, which began about 12,000 years ago.

One lesson of the Ice Age is that substantial effects that ripple across ecological zones and all of life were caused by what seem to us to be a numerically small change in average global temperature. That is because rainfall, winds, and the dynamics of all climate variables around the seasons were also altered. With an eye to the middle estimate for the climate sensitivity of 3°C, we should be wary of a future Earth that will be changed more than one might at first think upon hearing that one number. With CO_2 doubled, temperatures could go up by about 60 percent as much as they went down in the last ice age. This is one reason why the climatologist James Hansen keeps emphasizing that we are creating a "different planet."[2]

Dave Helps Turn a Wind Turbine

In 2008, Dave the carbon atom was riding a swift stream of air. He was in CO_2 again, this time in the arid western part of Texas, a place of furious heat spells, legendary droughts, and strong winds. The steady strong winds are one reason why fields of wind machines have sprouted there. More primitive versions were called windmills, but today's electricity-generating wind machines are known as turbines. And across the Texas plain, widely separated wind turbines catch what is a natural transformation of solar energy: the kinetic energy in moving flows of air.

Dave was in an air current about 200 feet above the ground and a few hundred yards away from the giant spinning blades of a turbine rated at a million watts of peak power. Cruising at 15 miles per hour on a typical winter day with mild mid-afternoon temperatures, Dave and his molecular neighbors suddenly began to slow down. The turning blades ahead, as they captured the energy, made Dave and his neighbors pile up like rush-hour commuters at a subway turnstile. Like the commuters, who get closer to one another in front of the turnstile, Dave and the other gas molecules, mostly oxygen and nitrogen, were squeezed slightly closer to one another as they lost speed. Energy was being transformed from the kinetic energy of velocity to the pressure energy of actual air pressure.

After crossing the plane of the spinning blades, which extracted energy from the pressure, Dave and his neighbors gradually sped up on the other side as they became again part of the overall flow of the wind. Some of Dave's kinetic energy thereby went into the turbine's generator, which sent electricity down wires inside the turbine's steel trunk, down to the ground, and into a power grid extending toward Houston and Dallas.

At the same time, Coalleen was dissolved in water, within a bicarbonate ion, in water that came from Lake Erie, the smallest of the Great Lakes. She, too, had begun to travel in a flow, not of wind but of water, in the Niagara River, which might have dumped her and her neighbors (mostly H_2O molecules) into Lake Ontario by way of Niagara Falls. Such was not to be Coalleen's fate, however. Upstream from the falls, at the great river's eastern bank, Coalleen flowed into a huge pipe that took her downward toward a water turbine.

As Coalleen flowed downward, she was pressurized by the weight of all the water from behind. This pressure impelled water through the plane of the blades of the fast-spinning water turbine. Coalleen did not actually touch the blades, but that did not matter; she participated in energy transfer at a place where the world's first large-scale commercialization of electricity had begun. She left the piping and was discharged to the gorge downstream of the falls.

At the same time, Icille, like Dave, was in the air within a CO_2 molecule. She had been in the air since 1994—statistically longer than the average, but not an uncommon amount of time. In western Pennsylvania, she got pulled into a plant's leaf and brought into its photosynthetic machinery. The plant was switchgrass, a tallish, lanky, fast-growing plant that is a candidate for biomass energy.

A Whirlwind Tour of Energy Systems

The wind, water, and biomass that carried our carbon atoms are all, in a real sense, forms of solar energy. The sun empowered the switchgrass leaf to bring Icille into an organic, high-energy molecule. The sun also evaporated and thus lifted the water, which perhaps originated as vapor from the Pacific Ocean at sea level and fell as rain into the higher ground around the Great Lakes, eventually to become the gravity-driven flow though the Lakes that carried Coalleen into the hydropower station near Niagara Falls, where her potential energy was converted to kinetic energy. And the sun powered Dave's momentum as part of the Texas wind, because it is the sun's torrent of parallel rays upon the round Earth that creates hot zones in the tropics and cool belts further north and south. This temperature difference (with others, such as those between ocean and land) sets in motion the swirling giant ribbons of air that are always seeking to restore energy imbalances created by the sun. Stop the sun and photosynthesis halts, the water cycle is closed down, the winds are braked.

These renewable, sun-derived forms of energy are held up as possibilities for the post-carbon world. No CO_2 at all comes from a water turbine or a wind turbine. Solar photovoltaic panels, which make electricity directly from the sun, are another candidate and are being deployed in increasing numbers. Critics might say that fossil fuels went into the manufacturing of these devices if they were built in factories that used electricity from coal-fired power plants. That's generally true today, but it is not germane. Wind turbines could be built in factories that would use energy from wind turbines.

Nuclear energy is like hydro, wind, and direct solar in that it neither uses a carbon-based fuel nor emits CO_2 as a waste gas. In nuclear fission, uranium atoms are split in a controlled manner into other types of atoms, most of which become radioactive wastes that must be kept away from groundwater, the air, and the hands of terrorists. I would be much more supportive of nuclear power if the world were full of perfectly peaceful people who would never dream of stealing radioactive materials for the purpose of building nuclear bombs or even "dirty bombs" that could spread radioisotopes across an urban landscape. Some experts, however, claim that we can handle substantial increases in the amounts of nuclear waste and keep them confined and safe.

Energy technologies of all forms have their distinctive negative effects on the environment. The impacts are often numerous. Nuclear fission, for instance, has the additional problem of a limited supply of the fissionable form of uranium called U-235, as the energy analyst Marty Hoffert has emphasized. Breeder reactors, which do not now exist on a commercial scale, would breed more fuel than they consume but would increase the danger of proliferation of nuclear materials significantly. Hydroelectric power, though it will continue to be important, is limited in its potential for additional development. For industrialized countries, it has essentially been tapped. With wind, solar photovoltaic, and biomass, a major concern is the enormous scale involved.

A recently published plan for deploying a mix of photovoltaic panels and solar concentrators in the southwestern United States was designed to fill 69 percent of projected electrical needs by 2050.[3] Carbon emissions would be cut by 75 percent by reducing the burning of coal. The solar arrays would occupy about 46,000 square miles (120,000 square kilometers) of land. That is only 1.5 percent of the land area of the contiguous United States, but it is 40 percent of the state of Arizona. As the authors of the plan show, deployment could be spread across the states of California, Nevada, and New Mexico, in addition to Arizona.

The area required to produce a certain amount of energy from fields of giant wind turbines will depend, as does the area for solar energy, on the resources available at specific locations. Because wind power increases as the cube of the wind velocity, siting is crucial. North and South Dakota have great potential for the extraction of wind power, and that is the kind of areal coverage that would be needed to make deep cuts to current and projected U.S. carbon emissions. On the plus side, when turbines are spread across agricultural land, about 90–95 percent of the land can still be farmed, with some farm roads doing double duty for maintenance of the turbines.

The outputs of energy from sun and wind are intermittent. Thus, if those energy sources are to become more than something like candles relative to the floodlights on the global energy stage, their outputs will have to be stored in some manner. The solar plan cited above would shunt direct current to other parts of the country where it is needed and where there are underground caverns in which to store the energy, in the form of compressed air, for generating alternating current when needed. Other plans call for producing hydrogen (from water) as a form of chemical energy storage. Energy storage on enormous scales is recognized as an outstanding engineering challenge.

We do know how to grow green plants. Thus, in addition to capturing the sun's energy indirectly via wind or by direct conversion to electricity with solar panels, we could harvest it, literally, in the form of biomass—the carbon-based molecules of plants. We could burn the plants. Fire from biomass combustion can make steam, the steam can turn a turbine, and the turbine can produce electricity. Wood-burning electrical plants currently exist. In addition, biomass can be converted into liquid fuels, such as ethanol or biodiesel. With biomass, because it is relatively stable, you automatically get energy storage, just as with fossil fuels. Could renewable biomass, or bioenergy, be significantly expanded in scope?

The use of biomass does produce CO_2. But remember, carbon was taken from the atmosphere to grow the plant. This is like the case of Dave in a molecule of alcohol (ethanol), soon to be emitted as CO_2 from a human breath. The exhalation will produce no net emission of CO_2, because that bodily emission is only one part of a cycle. With biomass, we are talking about replacing our fossil-fuel servants' emissions of CO_2, which are true net additions to the atmosphere, with cyclic emissions from biomass energy servants. Therefore, in the language of the energy analysts, those bioenergy servants would be carbon neutral.

Biomass energy, like wind and solar energy, requires large areas of land. In the United States, the production of relatively small amounts of ethanol from corn and the anticipated expansion of that industry has already driven up both the price of corn (also used as feed for meat animals) and the price of land on which to grow corn. There is no way to avoid the reality that the expansion of bioenergy will compete with land for agriculture. How much land is required? It would take an area of land about equal to that currently used by all agriculture to satisfy the world's energy needs.[4]

We can see the magnitude of the potential of bioenergy in plain global terms by looking at the annual oscillation of CO_2 at Mauna Loa (figure 5.2). The data from Mauna Loa have been shown to provide a very good proxy for the CO_2 of the northern hemisphere average. The annual plunge in CO_2 from its peak values in May to its lows in either September or October, as the air's carbon is sucked down by photosynthesis during the summer's growth of land plants, overcomes the increase from the steady CO_2 input of fossil fuels by a factor of 6. Nature's fluxes are larger than the fossil-fuel pollution. The difference is that nature's fluxes occur in cycles, whereas the industrial flux is a steady addition. But the fact that global photosynthesis and respiration are larger fluxes of carbon than the fossil-fuel emissions shows that nature has the potential to allow us to be carbon neutral if we can take the products of photosynthesis and "respire" them with oxygen as fuel to generate electricity (as one option). Advocates of a different tactic, "global gardening," submit that it would be possible to bring CO_2 levels down almost immediately by managing forests and soils in such a way that they would sequester carbon naturally.[5] Either way, the areas involved are huge. To fend off a nasty competition that drives up food prices (already happening) and to avoid clearing new land (which releases carbon dioxide), it will be crucial, in my opinion, for agronomists to create field technologies and high-tech plants that are at least a factor of 2 more efficient in converting solar energy into carbon-containing biomass.

Some energy analysts warn against the renewable "solar" forms of distributed energy because of these large areas. Jesse Ausubel of Rockefeller University boldly states that "renewables are not green." The Gaia theorist James Lovelock despairs at calls to deploy 100,000 wind turbines across the English countryside. Ausubel and Lovelock both support expanding the use of nuclear fission.[6] Other analysts foresee the possibility of continuing to burn fossil fuels without releasing CO_2. At first that seems impossible, because if you use a fossil fuel without generating CO_2 you don't get much energy, except in the case of natural gas, whose molecules contain significant amounts of hydrogen, which can be burned. (Once again, oil has far less hydrogen, relative to carbon, than natural gas, and coal has even less. Thus, per unit energy generated, coal is the worst fossil fuel in terms of CO_2 emissions.)

So far we have seen two general categories of options in the quest for large-scale sources of energy that have no net emissions of CO_2. One category includes hydro, wind, solar, and nuclear fission, which in principle emit no CO_2 at all. The second includes carbon-based fuels made from biomass, such as fuel wood (direct burning of the biomass), bioethanol, and biodiesel. Fuels in this category do emit CO_2, but they draw their carbon from the atmosphere and thus, theoretically, have zero net emissions. There is a third category, as I have hinted. It includes ways of burning fossil fuels, creating CO_2 as a waste by-product, but, instead of letting the gas go into the air, capturing it and burying it somehow and somewhere, thereby preventing its release. Technologies in the third category go by the synonymous names *carbon sequestration* and *carbon capture and storage*.[7]

Sequestering the emissions from the 250 million motor vehicles in the United States alone would be nearly impossible, so proposals for carbon sequestration have focused on coal-fired power plants, which are large enough for economies of scale to kick in. "Scrubbers" in power plants would remove the CO_2 generated in advanced combustion chambers. The CO_2 could be piped into the deep ocean, piped underground into old coal seams or into deep, unusable coal seams, piped into depleted oil wells where the injections could help release more oil, or piped into deep underground saline aquifers that would never be used for drinking water or for irrigation. Right now sequestration is performed on several million tons of carbon a year. Stepping that up to what experts estimate could be as much as one-third of global emissions by 2050 would require an amplification of more than a thousandfold.[8] Furthermore, most of the proposals are as yet untested. It is worrisome, for example, what earthquakes might do to the carbon put into the deep aquifers on land, or what the addition of concentrated plumes of CO_2 sent down into the ocean would do to marine life in the vicinity being acidified. The masses required to be moved in sequestration schemes are enormous, but then, it could be argued in support, so are the masses of fossil fuels already being transported daily around the planet. In the category of sequestration we should also include proposals for cleansing the atmosphere of CO_2, perhaps by erecting football-field-size towers of absorbent chemical solutions, perhaps followed by precipitating solid bricks of calcium carbonate for use in building projects.[9]

Research is also being conducted on nuclear fusion. This energy technology would imitate the energy production of the sun, where small atoms of hydrogen are fused into larger ones of helium, releasing energy in the conversion. High temperatures and pressures are required, which would destroy materials we currently know, and thus research focuses on confining the fusion reactions within magnetic "bottles." Radioactive wastes would be generated as the surrounding walls are hit by radiation, but the amounts would be small relative to the inherent wastes from current nuclear fission. The trouble is, fusion in the most expensive experiments has only been maintained for fractions of a second. To some, fusion seems to be the ultimate answer, if (and that's a big if) it can be done. But what if it can't be done, and in time? Some analysts claim that even the materials outside the magnetic bottles will never be able to withstand the fusion reactions long enough to be economically viable.

Many smart and skilled experts have visions of more futuristic non-CO_2-emitting energy systems. One vision involves giant orbiting arrays of solar panels, each square meter of which would be bathed in full sunshine 24 hours a day. The energy could be beamed down to the surface using microwaves or other frequencies. In another vision, small fusion plants would be surrounded with materials designed to be irradiated and thereby bred into fissionable elements subsequently sent to advanced nuclear fission plants. Ways to harness tidal power, geothermal power, or even Earth's magnetic field are either in small-scale use (often locally important) or in the fantasy stage. As the energy analyst Marty Hoffert has shown, there is no obvious solution out there. An all-out research effort, with many seeds planted, should be started, so we can see what deserves to bloom (in terms of economic and environmental feasibility) and to be deployed on a large scale.

There are numerous initiatives, aimed at improving energy efficiency and energy conservation, that would provide "negawatts" (a term championed by Amory Lovins, an energy analyst at the Rocky Mountain Institute). The compact fluorescent bulb and the LED light are "poster children" for the negawatt paradigm.

What if we could travel in lightweight carbon-fiber hybrid cars that would get 100 miles per gallon? What if everything we use that requires energy could be redesigned with the highest possible efficiency? This is the vision. What I like about an all-out effort to improve efficiency is that everyone benefits. With negawatts, no country has inordinate control over a resource, as Saudi Arabia now has over oil.

Figure 7.6 compared the numbers for the world's increases in the annual generation of wealth (in terms of gross world product) and energy consumption from 1970 to 2000. To increase wealth in 2000 by a factor of 2.6 from its 1970 value took only 1.8 times as much energy. The world became better than 40 percent more efficient at generating wealth from a given unit of energy. To clarify, the world required 40 percent less energy in 2000 then it would have used if it had generated its year-2000 wealth using the energy technologies and other infrastructure of 1970.

Note that the Central Trend projection for emissions described at the end of chapter 8 assumed a continuation of historical trends for energy conservation and efficiency as well as structural changes in the sectors of the economy, such that approximately twice the wealth will be generated per unit of energy in the year 2050 relative to 2000. Thus, a strong improvement in the use of energy to make wealth already has been assumed, on the basis of historical trends. Of course, there will be limits to these trends, such as thermodynamic limits to energy transformations and engineering limits associated with economic tradeoffs, and exploring these limits provides a rich field for debate among experts.

Projecting the future of energy is rife with competing ideas. One group of competing mutant ideas goes under the rubric of geoengineering. Theorists in this camp say that we might have to counter the eventual future warming, should it become dangerous, by treating Earth as we now treat buildings: as a problem of HVAC (heating, ventilating, and air-conditioning). Has it become too hot? Then we'll cool things off, perhaps by deliberately dispersing reflective sulfur aerosols in the stratosphere or by deploying arrays of millions of tiny auto-orienting mirrors way out in space. Both schemes would aim to deflect a precise percentage of the sun away from Earth. By offsetting the increased greenhouse effect with the right amount of cooling via geo-engineering, people in the future could return the climate to what it was back in the twentieth century (or any climate they wanted, even), and very likely for a lot less money than it would take to rebuild the world's energy systems.

Even if such schemes to cool Earth were to work, the CO_2 would still be elevated. As CO_2 increased in the oceans, we would then be confronted by ocean acidification, a problem independent of CO_2's greenhouse effect. CO_2 dissolved in water creates a mild acid (carbonic acid). Measurements have already shown that the ocean has dropped by about a tenth of a pH point because of its net absorption of some of the excess CO_2 from the atmosphere. Acidification will continue as CO_2 rises. Most obviously at risk are organisms that precipitate certain types of calcium carbonate shells, such as corals. Predictions of ocean acidification become especially serious after 2050, but some marine biologists foresee serious problems even before then. In a recent paper in the journal *Science*, a large international team of marine biologists and biogeochemists paint a bleak picture for the future of coral reefs.[10]

Many energy analysts predict that no singular solution will emerge from the cultural evolutionary field of variations and the selective market forces of supply and demand. They see as the likely answer the very existence of multiple solutions: for example, perhaps a Rhode Island-size area of solar panels (not 40 percent of Arizona) along with an Indiana-size area of wind turbines (not one the size of North Dakota). They foresee some increase in nuclear power stations, some future carbon capture and sequestration in advanced coal power plants, some biomass energy farms for both direct burning and also for liquid fuels, and a lot of high-efficiency vehicles, along with an all-out redesigning of the entire gamut of energy-eating technologies.

Stephen Pacala and Robert Socolow have developed a scenario to hold the world's CO_2 emissions constant all the way to 2050, starting as soon as possible.[11] Their program conceives of a multi-wedge attack in which each wedge grows to replace by 2050 what would have been an annual billion tons of carbon released as CO_2 from fossil-fuel combustion. Thus, instead of 14 billion tons of carbon released in 2050, which is approximately the IPCC's projection for its "business as usual" scenario and close to the Central Trend projection in figure 8.4, the world would emit only 7 billion tons of carbon. Seven of the billion-ton displacement wedges would have to be in place by 2050. The wedges include hybrid cars, nuclear power plants, and wind turbines. Many other energy analysts have also championed scenarios that promote multiple solutions.

Sounding more of a warning regarding the difficulty of bringing about the necessary technological changes, Marty Hoffert of New York University calls for nothing less than an energy revolution. Hoffert and others[12] have made analogies to modern history's most stupendous recent scientific and engineering leaps—for example, the Manhattan Project and the Apollo Project. Hoffert emphasizes that extraordinary gearing up is possible. He points out that from 1939 to 1945 the United States increased its annual production of aircraft by a factor of 50. We need what might be named the Prometheus Project, after the hero of Greek myth who stole fire from the gods.

Three Scenarios for the Future of Emissions

I once saw a cartoon that depicted the dome of the U.S. Capitol poking up just above sea level and two congressmen mounted on inflatable flotation devices, one with the head of a donkey (symbol of the Democratic Party) and one with the head of an elephant (symbol of the Republicans). One congressman says "But, if we'd done something about th' greenhouse effect, we wouldn't have this neat swimmin' hole."[13] If we aren't fearful of a specific future (who knows whether the United States will even have a congress when the ice sheets melt), we can be fearful of the gambles we are taking with the state of the planet through collective participation in the global industrial growth automaton. But what would be gained by reducing CO_2 emissions at a specific rate?

I am going to show you results from three scenarios, which should serve to frame the problem. Furthermore, my focus at first will be up to the year 2050, since that is within shooting distance for our minds, given that many us now reading books on the carbon cycle or global warming will be alive then or almost make it. Further time projections will make more sense when the changes in CO_2 and global temperature are first projected out to 2050 in three scenarios and the results compared and understood.

The medium projection is the Central Trend scenario, in which CO_2 emissions from fossil fuels in 2050 are about twice what they were in 2000. To evaluate what effects on atmospheric CO_2 would be from alternative emission scenarios, it would be appropriate to develop both a lower one and a higher one.

Pacala, Socolow, Hoffert, and others have been discussing a particular low scenario that we might call "Constant Fossil Fuel." It would be extraordinary, in my opinion, if this low-emission plan could happen, because the current thrust of the GIGA is tied so intimately to the expansion of fossil fuels. I am not going to even try to deal with the costs of the Constant Fossil Fuel scenario; I want only to focus on the results for atmospheric CO_2 were this scenario to be followed.[14] There are two thrusts of innovation that could be combined to allow this to happen. First, improvements in energy efficiency and conservation could be stepped up above the historical rate. Second, we could bring on truly substantial amounts of sources of energy that have no net emissions of CO_2. Any of those described above could work in principle.

What I am going to do for the scenario of Constant Fossil Fuel is hold emissions constant starting in the year 2010. Again, I am not going to detail how the thrusts for technical innovations and regulations might be combined with economics to achieve this low emissions path; I merely want to posit the path as a "what if." What if the CO_2 emissions could be held constant at projected 2010 rates until 2050? How much lower would the atmospheric CO_2 be, relative to the level reached by the Central Trend? Given that the global emissions rate is now climbing each year, and given the long lead times for drastic changes in technologies (a power plant is built to last 50 years), this scenario is probably a radical lower bound on what is possible—a fantasy really, but, as I hope you'll see, an instructive one.

We might consider, as a third scenario, a radical upper bound as well. Recall the fact that Central Trend-2050, which projects current trends for gross world product, energy, and CO_2 emissions, already has built in a twofold improvement in the generation of wealth from a given unit of CO_2 emissions. This improvement is potentially justified by many technologies that can in fact be deployed, such as wind and nuclear, as well as hybrid cars and other numerous possibilities in energy conservation and efficiency everywhere. But what if we don't achieve even that? What if the ratio of CO_2 to GWP does not decline? (Indeed, a recent study shows that for early years of this century the historical decline has stopped.[15]) China is constructing coal-burning power plants at a torrid pace. Coal is the worst fossil fuel in terms of its carbon emissions per unit energy. China and India have large amounts of coal, and it would be foolish to assume that they won't use it, just as other countries during economic development have used what they have been given by the luck of the geological roulette wheels from millions of years ago. The United States has been called "the Middle East of coal," and it is possible that when the looming "long emergency"[16] of an oil shortage arrives the United States will begin converting its substantial geological endowment of coal into liquid fuels for its transportation needs. Without carbon sequestration, this could take the emissions higher than the Central Trend-2050 path. Thus, for an upper bound, I posit a scenario in which GWP grows at the rate of the current trend and CO_2 emissions also grow perfectly correlated and at that same rate. It projects the GWP growth locked to the technological capability of the year 2010. This scenario will therefore be called "Frozen-2010 Technology."

We now have three distinct scenarios that differ dramatically in their CO_2 emissions: the medium Central Trend-2050 scenario, the low Constant Fossil Fuel scenario, and the high Frozen-2010 Technology scenario (figure 9.1). Again, I am not dealing with costs (not my realm of expertise, and debatable even among economists). I am only interested in examining what the scenarios mean for the future of atmospheric CO_2 and thus the greenhouse effect caused by CO_2. To that I now turn.

CO_2 Levels in 2050 According to the Three Scenarios

How do the three scenarios play out in producing higher CO_2 levels in the year 2050? And what do those higher levels mean for global temperatures?

I am going to show the results in a graph. But before I do that, allow me to make a few explanatory remarks.

I used a slightly different airborne fraction for each scenario, to take into account the fact that for larger amounts of emissions, the airborne fraction is itself expected to be larger, as shown by modeling results.[17] I have also verified the results using a more complicated model, and the results are consistent with modeling projections run by others, using more sophisticated models, and I recognize that such models are re-

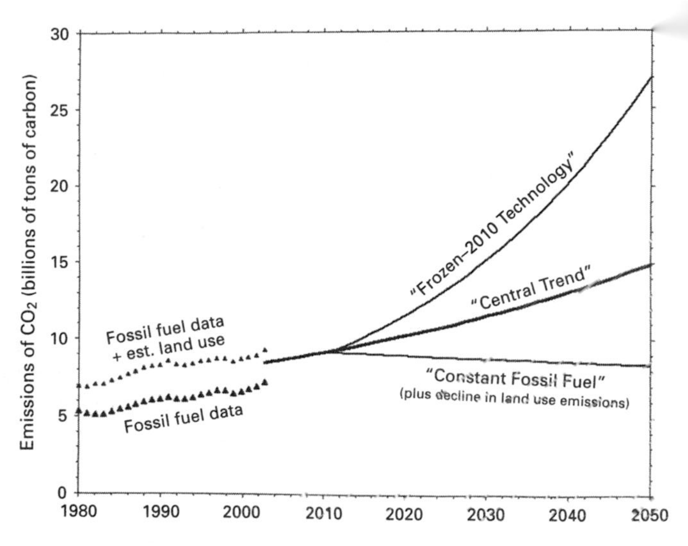

Figure 9.1 The three scenarios described in the text for the future of CO_2 emissions. Consistent with an average derived from a group of IPCC projections, I have added in a small amount of land-use emissions, which decline over time and are the same for all scenarios, but this is a minor detail relative to the dramatic spread generated by the different assumptions for the future of fossil-fuel emissions.

quired if one's goal is to explore unexpected degrees of feedbacks in the system.[18] But we have already seen that the airborne fraction has been remarkably constant.

The equilibrium temperatures will be shown on the right side of the graph, assuming that the equilibrium temperature rise will be 3°C at 560 ppm of CO_2, approximately the value for CO_2 doubling from its pre-industrial state. These temperature numbers are not the degrees of global average warming that would exist at the time any given CO_2 value is reached; they are the eventual degrees that would be obtained if each given CO_2 value were held constant long enough for the surfaces

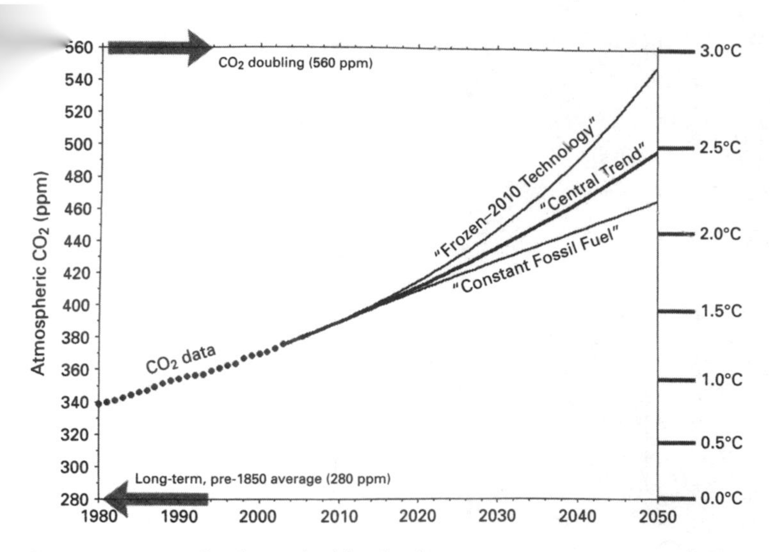

Figure 9.2 CO_2 levels reached by the three emissions scenarios in figure 9.1. The temperatures are for the equilibrium temperature increases from CO_2 forcing alone that would eventually be achieved for each CO_2 level after the ocean, with its thermal lag, reaches equilibrium. For the temperature scale I use the IPCC climate sensitivity of 3°C for CO_2 doubling.

of the oceans to warm up and come to at least near equilibrium with the atmosphere. The large mass of the oceans creates a thermal lag on climate change of several decades, for the largest fraction of any projected warming, but in fact some slight amount of the lag stretches out for a much longer period.

Enough preliminaries. Figure 9.2 shows the projections.

I will now elaborate on the finding that the differences in the CO_2 levels generated by the upper and lower scenarios, relative to those of Central Trend, are not remarkably large.

First, please compare the results for "Central Trend" and "Constant Fossil Fuel." What do these two quite different scenarios for emissions mean for the atmospheric level of CO_2 in the year 2050? Cen. Trend's CO_2 value for that year is 496 ppm, and that for Constant Fos sil Fuel is 464 ppm. Thus, if the world is able to take what seems the strategically very different pathway of Constant Fossil Fuel, the saving is slightly more than 30 ppm. Put in the perspective that the overall rise started in 1850 at a pre-industrial value of 280 ppm, the saving of 30 ppm relative to an overall increase of about 200 ppm is on the order of 15 percent. This can be seen in the graph, because I have positioned 280 ppm as the "zero" line for the CO_2 levels.

Is that 30 ppm (or 15 percent) saving a lot, or a little? In one sense, it is a lot. It is the amount the atmosphere has gone up in the last 18 years. But my general reaction was one of dismay when I first saw this. Is the world going to embark on a crash course to shift its energy technologies to save 30 ppm in the year 2050?

In comparing the two scenarios in terms of the equilibrium global temperatures they produce, the saving is even slightly less. As you can see from the graph, the amount of global warming does not scale linearly with the atmospheric CO_2. The climate forcing from the greenhouse effect of the first 100-ppm increment above the pre-industrial state is more (by about 30 percent) than the climate forcing from the second 100-ppm increment. With the equilibrium temperature scale I have used,[19] the amount of warming avoided by going to the Constant Fossil Fuel scenario, relative to the Central Trend-2050 scenario, is 0.3°C. Should anyone advocate ways to put the world on a high-gear course to level off CO_2 emissions to the year 2050 to save 0.3°C (about 0.5°F)? It would all depend, of course, on the costs.

Some urge caution about bringing up uncertainties about costs because people could be scared away from the call to necessary action, and many expect economic benefits from redoing the energy infrastructure. The savings in avoided CO_2 and global warming, though small at st, will bring in the new technologies that will be hugely profitable, as the old fossil fuels, such as oil, go up in price and as more and more regulations are placed on the combustion of fossil fuels. But critics of that might say that it is one thing to replace light bulbs but another to retool and redesign everything that that requires and uses energy, and that will cost big money. Those same critics might admit that the installation of wind turbines (to take one example of an energy source that does not emit CO_2) is increasing at a high annual rate. That, however, is partly because of subsidies and partly because they are cheap now because as a total amount of energy supply they are small relative to the electricity-producing systems that use coal, oil, natural gas, hydro, and nuclear. The current wind (and solar) systems can be merged into the existing electrical grids. But if these renewable energy systems become a larger proportion of the total, their intermittency becomes a huge headache. It must be addressed by developing the capacity to store their generated energy during times of abundance, for use during the down times of the geophysical fluctuating supplies. And that is going to raise their costs substantially, especially because (as has been noted) the technologies for such huge storage capacities remain to be developed, proven, and made cost effective. Many experts are also talking about the fact that the entire electrical grid will need to be reconceptualized and redesigned in fundamental ways. If hydrogen is to become the storage, then we need an entire hydrogen infrastructure.

Another way to look at this situation is to suppose that the coming "long emergency" of an oil shortage will save us (if it doesn't sink us). I think it is almost a sure bet that an energy crunch from dwindling production of oil will arrive before the CO_2 warming causes general apprehension and fear to push global warming toward the top of the agenda for real political action for everyone in the world.

Some experts claim the oil production is peaking now; some say the peak is 10 years away, some say 20 years, some even say more.

But these time spans are short to a world economic system that sucks up more than 80 million barrels of oil a day with no replacements in sight. In comparison, the Arab oil embargo of the 1970s was relatively simple—OPEC merely had to be talked into opening its valves up (really downward) to the subterranean again. Peak oil is going to be closer to a situation of an exploding population of bacteria reaching the edge of the Petri dish. I don't mean the human population, rather the population of the oil guzzling segment of the energy servants being manufactured across the world to serve the public. The servants will run up against the walls of a limiting resource.

That might rev up the industrial market machinery to develop serious amounts of replacement energy sources. Now, if countries with large coal reserves (such as the United States) could refrain from liquefying coal without carbon sequestration and instead create an energy infrastructure that is carbon neutral or balanced, then solving the more immediate oil crisis might solve the longer-term greenhouse problem. But one never knows which way a crisis is going to go. Look at the ruins of ancient Rome.

Yet another assessment comes from some prominent climatologists who claim that the level of 450 ppm is worrisome. Indeed, this is starting to be certified by the international scientific community as the value not to exceed. At that amount, with its warming of about 1°C on top of the warming to which the world is already committed (but has not yet achieved because of the ocean thermal lag), the biosphere will enter territory that is unknown, in terms of our detailed knowledge of past climates, which at least help in forming expectations. That level, they say, could push the world into an unprecedented state with disastrous effects on coral reefs from ocean acidification, with major shifts in biodiversity and ecosystems from altered rain and temperature patterns, and with the potential for an equilibrium sea level that is several or more meters higher, when the ice sheets melt to achieve the new equilibrium state.[20] The level of 450 ppm is surpassed by all my scenarios by 2050. The only way that CO_2 levels will fall is by radical drops in emissions. Yes, it's agreed, there is little difference in 2050 between the Central Trend scenario and the Constant Fossil Fuel scenario. But if the Central Trend scenario is bad, so is the Constant Fossil Fuel scenario.

The Issue of Technological Inertia

Colleagues I approach with the results shown in figure 9.2 often have another response to the fact that the savings in 2050 between the medium Central Trend scenario and the low Constant Fossil Fuel scenario are small. To develop this line of reasoning, I want to first look back to the results of the high scenario of Frozen-2010 Technology. It produces a CO_2 level about 50 ppm above that of Central Trend. And its equilibrium temperature rise is about 0.4°C above the warming produced by Central Trend. So by ensuring that the world in the coming decades lowers the energy required to produce a given unit of economic wealth along the historical trends that have been followed for the last few decades, rather than creating the wealth of the future using year 2010 average technology, the world prevents 50 ppm of CO_2 and 0.4°C of warming.

Again, one could argue that that isn't much. But the savings are free, in a sense, from following the trend we have been on, which has come about naturally. Energy is expensive and industries gain by controlling costs. And regulations have helped somewhat. But the main point really has to do with the concept of technological inertia. If the energy technologies have not changed by 2050 (and, again, they will be forced to because of the looming oil crisis, but that doesn't mean that the new ones will emit less carbon, so please allow me to follow the logic), it is doubtful that they will change swiftly in the future after that. Power plants are built to last half a century or more. Factories set to manufacture certain kinds of cars will tend to keep manufacturing those cars. There is inertia in the GIGA, just as there is in an expanding biological population.

This concept of inertia is at the heart of the answer I hear when I solicit opinions about these issues from colleagues, many of whom have spent substantial portions of their professional lives in research about the carbon cycle and global warming. They agree that radically different emissions scenarios (at least within a conceivable range) produce relatively small differences in CO_2 and global temperature in 2050. But the big game is afterward, they point out, in what will happen between 2050 and 2100. Take a look at figures 9.1 and 9.2 and start projecting those curves into the second half of the century. It is then that huge differences among alternative paths will appear, differences that could mean a yea or a nay to the start of runaway melting of the ice sheets of Antarctica and Greenland, or a runaway amplification of worldwide weather disasters, or the switch in land ecosystems from a regime of absorbing CO_2 to one of emitting CO_2 (as the soils and tundra warm), or some other tipping point in the climate system. We know that tipping points have occurred in the past, shown by the glacial cycles, which prove that climate is sensitive and contains amplifying feedbacks.

So which emissions path the world follows up to 2050 might not by itself make much difference in 2050, but that path, due to interconnected inertias of economics, technology, and social factors, will diverge from its alternatives in a large way by 2100. For example, even if the world is on the Central Trend of rising emissions in 2050, that implies that fossil-fuel technologies are still being built then at an increasing rate. It means the GIGA will still be geared then toward increasing fossil-fuel combustion (in the absence of substantial carbon sequestration) and thus its momentum will most likely carry the world along the path of increasing emissions even further into the future. Thus, it appears that getting super-concerned right now requires getting concerned about the second half of the twenty-first century and about future generations.

On the other, more cautious hand, consider this: Although the Central Trend has increasing emissions, it still implies substantial technological changes relative to the path of Frozen Technology. Central Trend has only half the CO_2 emissions for the production of a given unit of wealth (as GWP) in 2050, relative to what that ratio was in the year 2000. That means substantial changes in the social infrastructure (services increase relative to manufacturing), in energy efficiency and conservation, and in new energy sources that do not emit CO_2. These changes are the ones that would be needed, though at even larger scales, to level off emissions and eventually bring them down during the second half of the century. And achieving the Central Trend itself might be no easy matter. But if historic trends continue, it seems to be a possible path, given that so many experts foresee the potentials to lower emissions from current technologies across a variety of fronts. Many of the required technological changes after 2050 will have been put into gear before 2050 by the Central Trend, and at that time science will know a lot more about the effects of CO_2 on climate, having had decades of careful monitoring and modeling of melting mountain glaciers (if any are left), retreating sea ice (the Arctic will likely be ice-free in summer), and the ice caps of Greenland and Antarctica, as well ecosystems, rainfall patterns, hurricanes, and droughts. So perhaps we shouldn't worry, too much, at least right now, other than ensuring that the improvements implied by the Central Trend come to pass. Should people (those in developed countries, at least) just say "Carry on"?

Not quite yet. All this discussion has been about the global amount of emissions. As I hope to show, another view comes into focus when we look at the situation on a regional scale. Though in this book I have often emphasized the global numbers, countries are still crucial because they are the units at which laws get enacted across an entire group of people with a certain per capita average something, such as for GDP, energy consumption, and CO_2 emissions.

The Obligations of the United States and the Other Large, High Per Capita Emitters

Here is what I think is not only probable but (I hope) acceptable: Relative to the year 2000, the world carbon emissions will double by year 2050, along with the desirable quadrupling of the gross world product, which occurs if the world continues the growth rate of the current GIGA. At that point, the average world citizen will have reached an economic level about equal to that of the average European or Japanese in the early years of the century. After 2050, the various emission scenarios diverge into a more dramatic spread of ever higher global temperatures, with varying degrees of uncertainty relative to the spread. To avoid the more extreme marches into climate regimes that would likely be disturbingly unpredictable and unacceptable, it will be necessary to level off global emissions in the Central Trend scenario after 2050, and then bring them down toward the end of century, by deploying and advancing non-CO_2-emitting energy systems, energy efficiency, and conservation. Noteworthy in this overall vision is the fact that *substantial amounts* of exactly such helpful technologies would have to be in place in 2050 just to follow the path of the Central Trend itself. Such advanced technologies are required in the Central Trend and are the very developments that avoid the degree of fossil-fuel technological inertia that would otherwise still be in place in 2050 were the world to steam ahead to there using the high-emissions track of the Frozen-2010 Technology scenario.

But the devil of acceptance of the possible doubling of world emissions will lie in the details. For example, the global doubling could come about simply by doubling emissions from every nation. But this case would be disastrous to an environmental geopolitics that will require among its countries ever more coordinated negotiations about their carbon emissions. Clearly, the grounding for coordination is some converging degree of equity of emissions on a per capita basis, especially across the boards of the big national emitters. Items of agreement in the Kyoto Protocol already admit a principle of approximate equity (or at least the avoidance of gross, blatant inequity). The countries that have large per capita emissions are the ones responsible for the global problem and thus are the ones that must cut back at first. This basic realization obviously should not change. Without its further development, no advanced coordination will be possible. And without coordination in emissions, the world's countries in 2050 will be stuck deep in a state of profound economic and moral tension. We are already sensing the first roar of the possible coming avalanche of the economic and moral issues as well as a precarious, fearful sense about the future climate. Cooperation is dead if the avalanche that buries many of the innocent is perceived as being caused by just a few high-flying skiers.

Thus, although global emissions will rise in the coming decades, it is of vital importance to start bringing down the emissions in some specific places.

Surely the United States stands out as the singular example of a high-flyer, with its relatively large population and huge per capita emissions. In 2006, emissions from the United States were 5.3 tons of carbon in the form of CO_2 per person. The world average was 1.2 tons of carbon.

Now consider the Central Trend projection for 2050, in which annual world CO_2 emissions increase to 15 billion tons of carbon and world population reaches 8 or 9 billion people (with hopes held out for 8). The global per capita emissions would then be between 1.7 and 1.9 tons of carbon per year. That means that even if the United States were to level off its total annual emissions until 2050 (a huge "if" right there, given that its emissions have been rising annually), and assuming a generously large population increase of 33 percent (from 300 million to 400 million people), then in 2050 the U.S. per capita emissions would still be more than double the world per capita emissions at that time.

In this situation, the United States would then continue as the supreme example of the maxim of "wealth through carbon emissions," which implies that the technology to power its wealth would be still anchored to fossil-fuel combustion at per capita rates more than double the world average (again, with the assumption of level U.S. emissions to 2050). The United States would still be the paradigm for the way to link GDP, energy consumption, and CO_2 emissions, no matter what the consequences in the second half of the century for the atmosphere, for ocean acidification, and for climate with all the dice-tossing ramifications of global warming for ecosystems, agriculture, the urban environment, and human health and prosperity.

The United States is not alone in having high per capita emissions. Without getting into population projections, we can point a finger at all places with large populations that already have per capita emissions that are larger than the world per capita emissions projected for the year 2050 (from the published data for all the world's countries for 2004, and I will exclude as inconsequential to the big picture a number of small countries with high per capita emissions, which occur in the Middle East and in the Caribbean). Here are some at the end of that finger, with their emissions in parentheses, in annual tons per capita of carbon from fossil-fuel CO_2. Remember, for comparison, in the calculation just cited the Central Trend projection for 2050 has a global average emission rate of slightly less than 2 tons per person per year. But in year 2004 a number of countries were already above that: Canada (5.5), Australia (4.4), Saudi Arabia (3.7), Czech Republic (3.7), Taiwan (2.9), Russia (2.9), Japan (2.7), South Africa (2.7), Republic of Korea (2.6), and more.

The European Union deserves a separate discussion. Its countries vary in their per capita emissions, from the high of Norway (5.2) to the much lower France (1.6). But considered as a single entity, the EU's per capita emissions were 2.0 tons of carbon per person per year (for 2006, for the 15 pre-2004 members of the EU)—just slightly above the projected world average for 2050. Thus, the EU, if it were able to grow economically and at the same time hold its per capita emissions constant up to year 2050, could serve as a model of a highly developed region with only average world per capita emissions in 2050. The EU could hopefully start to bring down emissions after 2050, or even earlier, by continued development and deployment of the technologies that presumably would be phased in during an interval of continuous economic growth with constant emissions in the upcoming few decades. Even more encouraging, the EU is developing plans to reduce emissions 20 percent by 2020, which would likely bring it below the world per capita average projected for 2050.[21]

I have said that if the United States continues to have very high per capita emissions in 2050, which today are more than 4 times the world average, it would still serve as the example of how to gain wealth through fossil-fuel combustion. Depending on what kind of global ethics regarding waste carbon are in place by that year, the United States could become the prime example of how not to be, the actual devil in those details, the carbon-dependent "stink bug" of the world's countries. Of course, international opinion can only shift the national policies so far. But consider this path for the United States, one which would ensure a fertile field for international cooperation: to aim at lowering emissions to a per capita level close to the Central Trend's projected world average by 2050.

We need to assume economic growth for the United States, too, of course. No politician could run on a platform of holding things economically steady. Steady would be perceived as stagnant. Some of the millionaires in Silicon Valley look with envy on those with $10 million, who envy those with $100 million, who envy the billionaires.[22] Instead, what about positing a society where people are creative, producing stories to instruct and entertain, exploring personal growth detached from material growth because they have enough (speak for yourself, bud!). They could enjoy the world's great literature (aiming for wisdom, and that sounds good), drive smaller vehicles (possibly acceptable), and maybe live in smaller dwellings (wait, that's enough!). Some of my remarks in parentheses are from the little voice within me that says a reduction in material goals in exchange for more time for other values will never happen.

The coupling of economic growth with a reduction in emissions is a paradigm that will likely be required for the world as a whole during the second half of this century. But this very coupling (again, to reach some state of approximate equity in per capita carbon waste) is required for the United States very soon. Not then. Basically, now. Here is a tremendous opportunity for the United States to lead the way into what will eventually become the new global paradigm.

To couple a rising economy with falling emissions the United States will need a mix of the two major categories of technical strategy: (1) energy efficiency and conservation and (2) non-CO_2-emitting energy sources, probably from the list above. The United States has fantastic engineers. It is the envy of the world in engineering and invention. The

United States should be able to make advances across both categories. Maybe that will even include some carbon sequestration. There is funding going on for all these research fronts right now, for wind, solar, sequestration, nuclear, and advanced fossil fuels. The funding is probably skewed toward fossil fuels. But the money is there. The money, for example, that has been essentially committed to the Iraq war, a trillion or more dollars projected in long-term costs, could have constructed a new energy infrastructure capable of eliminating U.S. dependence on foreign oil and reducing CO_2 emissions by 75 percent from projected 2050 levels.[23]

A path in which all countries increase their emissions will lead to a world in 2050 that will be a continuation of the present trend, not just globally but also at the level of individual countries. What is needed as a model for the second half of the century is an example *now* of a large, complex, highly developed nation that continues economic growth with reducing CO_2 emissions. It is the trend of increasing emissions within all countries that is the real trend of danger to the state of the world in 2050, much more so than the possible paths of the summed global emissions. If 2050 is reached in which the poorer people of the world feel that the rich countries have gambled the climate by spinning a roulette wheel like gambling addicts, the rampant unfairness will create highly uncertain consequences for geopolitical cooperation. The poorer countries with rising economic muscles will then just continue to bid on fossil fuels, too. For if those fuels are still in increasing use in the developed countries, the implication is that the global economic system will have kept them nailed down as the fuels of choice.

In short, though the world will likely have globally increasing emissions, the United States should establish the goal of becoming an even wealthier society with substantially reduced CO_2 emissions.[24]

10

The Ultimate Fates of Our Carbon Atoms

When Dave the carbon atom popped out of the ocean and back into the atmosphere in the year 3279, the earliest decades of concerns about the rising levels of CO_2 and about the supercharged greenhouse effect were even further back in the past to the people of the thirty-third century than the European Medieval Ages, with their soaring cathedrals, armored knights, Thomistic proofs of God, and feudal castles, were to people of the twenty-first. But how those concerns had been met—with care, dismissal, ignorance, ingenuity, or conflict—had a lot to do with the flourishing or not of life in the biosphere and the sufferings and joys that filled it for at least the early centuries of that period of 1,267 years in which Dave had been riding the molasses-like currents of frigid water in the utterly darkest and deepest parts of the world ocean.

During his private millennium of submergence, Dave had been absent from prancing in any of the well-lit, fast-circulating, surface pools of carbon. He had not been in the atmosphere, absorbing and re-radiating infrared rays as a participant in Earth's greenhouse warmth. He had never even occasionally passed through the blades of the wind turbines that were being erected in ever-bigger arrays. He had never been embodied into a barley plant, or a grape, to end up as alcohol in beer or wine, perhaps drunk over dinner by international treaty makers or national legislators as they debated with more and more anxious vigor what to do about the rising CO_2. So Dave can not tell us anything about those crucial early decades of the twenty-first century.

Even if he had been around, he could not tell us, for he is a mere atom of carbon—six protons and six neutrons in a nuclear ball 100 trillion times as dense as rock, surrounded by six electrons in electrified, quantum-mechanical orbital clouds, at least four of which are able to forge bonds with electrons of other atoms, thereby giving the carbon atom, of all atoms, the highest versatility in creating a huge number of species of different molecules, including all those at the dynamical core of living things.

Methaniel, Oiliver, Coalleen, and Icille were there in the most active surface carbon pools during the crucial decades, nimbly awhirl in air, plants, bacteria, and other creatures, in soil, and in the surface ocean. They went through the standard rounds of adventures that carbon atoms do as they pass into, spend time, and then leave the most active biosphere bowls. Oh, yes, in 2015, Oiliver did have one noteworthy escapade. He was photosynthetically incorporated into first a leaf and then a stem of a genetically modified, nitrogen-fixing, experimental tree, targeted to grow like lightning and produce biomass yields double those of even high-yielding crops. The tree was sacrificed in 2019, dried and weighed as data, then ground up and pressed into pellets that were burned in a prototype of an advanced catalytic micro-powerplant, again for data. He reentered the atmosphere that year.

In 2020, Methaniel meandered into an even more unusual pathway. The lifetime of a carbon atom in the biosphere is usually about 100,000 years. That is only its average stay. For example, although in the early twenty-first century Dave had already been around for 32,000 years, for an average carbon atom his travels would be only one-third over. This specific residence length of an average carbon atom in the biosphere's cycles comes about not from any atomic analogy to an animal's or a plant's life span with its maturation and senescence. The carbon atom's "life span" derives from the statistics of the strands that pass among the biosphere's bowls. Very few strands went down out of the biosphere altogether and into rock until humans began engineering additional downward strands, one of which took Methaniel out.

A respectably bold experiment in carbon sequestration began in 2020 near a coal-fired power plant in Kentucky. Methaniel happened to be in the supply air drawn into the fluidized-bed combustion chamber of the operation, where inside the air's CO_2 concentration was boosted nearly a thousandfold from the gas waste as coal was consumed in the inferno. Most CO_2 molecules, including Methaniel (even though he was not waste at the time), were chemically captured and shunted down a pressurized pipe leading a half-mile underground into a saline aquifer. The sequestration engineers, members of a newly formed profession, estimated that the molecules would be safely ensconced for at least several thousand years. As the acidity level dropped along the percolating path of the aquifer, Methaniel, who had been moving at a snail's pace, precipitated out along the crack line of a silicate mineral and into a tiny crystal of calcium carbonate.

Dave, as already mentioned, except to provide the year as 2012, at least did stay in the biosphere but passed into the deep ocean by the following route. Having left my mouth during an exhalation the day after I drank that beer (yes, I was the drinker), Dave wafted around the globe for several years. Then he wriggled across the air's interface and into seawater in the Weddell Sea along the Antarctic coast. Within a bicarbonate ion he was just chilling out as winter came and the sea surface froze into a thick skin of ice, leaving Dave underneath, in water now saltier and heavier. Gobs of this dense water started to plunge downward, sinking and pulling other cold, salty water along in a dense plume. A year later, Dave had sunk to within a skyscraper's height of the bottom and began his 1,000-plus-year stay in what oceanographers on the surface call the Antarctic Bottom Water, the coldest, densest water mass in the world ocean.

Along with Dave in the downward plunging plume were other carbon atoms within bicarbonate ions, as well as some within carbonate ions, and some of all these came from CO_2 from every exhalation that I had ever made up to then. And from every exhalation that you had ever made.

The Carbon Challenge of the Early Twenty-First Century

In the early twenty-first century, people faced a major conundrum about carbon. The inequities in the per capita emissions across countries, and the links among GDP, energy, and CO_2 emissions made it painfully difficult to know what to do. It was easiest to let the Global Industrial Growth Automaton roll onward, and hope that politics, science, and engineering would solve any problems using cheapest short-term solutions before they might grow into slow emergencies or catastrophes. The beginning efforts toward developing a financial marketplace for trading permits on carbon emissions were well-intentioned and necessary first baby steps in a long journey of change, but were having little impact on global emissions. People in industrialized countries were comfortable. Those in developing countries strove for more comforts by joining the GIGA. So as the world's highway driving went up, to note just one form of increasing consumption, the world was driven deeper into the carbon conundrum. The challenge grew accordingly.

There would be no smashing collision into a brick wall. The conundrum came like heading off a paved road that had been easy to drive before plunging into a thicket. The brambles—climate problems and the difficulties of seeing a way out—began as merely annoying vegetation that impeded vision but then morphed into tangled webs of barbed wire. Global climate change came on slowly.

In a fast emergency people must respond. But in a slow emergency things are not so clear. Do you stop and reconnoiter, perhaps back up a bit? Or just push ahead and hope the way clears?

The math of the carbon cycle shows the slowness. The facts are these: Various radically different CO_2 emission scenarios, in which the upper and lower two ones are more fantasies that bound the probable reality of a path close to that of the Central Trend, do not make that much difference even in 2050, unless there is some kind of tipping point, like a fever that can go up: 100°F, 101°F, and so forth, gradually getting more serious but then when it reaches 104°F or 105°F a grave danger zone is suddenly upon the sick body, a point that surpasses the body's ability to regulate. The CO_2 is rising. The global temperature is, too. We could look closely at the differences in the projections, and attempt to split hairs to discern if a tipping point exists on one side or another of a difference. But from the viewpoint of the future, to those in our time it appeared that the entire game was really 50–100 years down the road. It did seem clear that the more the technologies of 2050 depended on fossil fuels, the more the global situation would engender still additional amounts of those technologies.

And hey, said some, why be so downtrodden with negativity? Doesn't one slogan that some derive from the situation of global warming claim that some will lose and some will win? For instance, amplification of the warming in the high latitudes is a virtual certainty, and there is talk that the countries that surround the Arctic Ocean will soon need to sign a treaty to divvy up the anticipated rewards that will come when that ocean opens up as the sea ice shrinks back year by year. The Russians have already sent a nuclear submarine to plant a platinum flag into the sea floor beneath the North Pole, claiming that Russian continental shelf goes out that far. Locally, the changes will be scattered and as yet highly unpredictable. One could embrace the argument, however blinkered to the potential for havoc to ecosystems and threats from sea level, that globally there is likely to be more total rainfall, and that could mean more plant productivity, more lushness, even bigger storms to admire for the seekers of the sublime in the powers of nature. All in all, as the biosphere goes into its brave newly warmed world, times will be interesting, and hopefully not just in the sense of the Chinese proverb.

Given the certainty of change and the uncertainty about the nature or speed of change, what else could people do but wait and watch as the losing and winning tickets were meted out by the gambling game of global warming? Human behavior had evolved to satisfy very real material hungers, for that was how the young could be safely raised. So everyone stayed harnessed to the Global Industrial Growth Automaton and hoped for the best.

Change will only happen when people demand it, and that tends to occur when negative emotions grow into powerful motivators.[1]

Fear is a negative, motivating emotion. Uncertainty in the weather has traditionally been one of the reasons people call upon a god or gods for assistance. Of course we have to be wary (fearful?) of being manipulated into unreasonable degrees of fear, which, for example, can lead countries into irrational wars. But certainly one message from paleoclimate studies is that the biosphere is sensitive and contains complex feedbacks that are far from fully being understood. The unknown is fearful, and going into climate change is heading into the unknown, like slow but inexorable aging, with its increasing odds of runaway disease.

Anger is another motivating emotion. One source of anger comes from the perception that the actions of others are endangering me and loved ones, a circle of concern that for many extends outward to embrace other species as well. There are many nuances to the arousal of anger. For example, there is anger at some governments and energy companies for acting to dissuade citizens of the reality of global warming.

Many people in these days of the carbon challenge have another, milder kind of fear, which is insidious and grades into the emotion of sadness and even disgust. It is an uneasy type of fear, always there in discussions, playing in the background like a slow dirge, a lament about our violating some sacred compact with the planet, not such that we should be terrified of the vengeance of a Kali or Gaia, but grounded in the recognition that all sorts of complicated transformations will come about merely from allowing our fossil-fuel energy servants to burden the air with their freely vented waste gases. Our wastes will be in our faces, in the faces of the wildlife, in the world's vegetation, as well as in the air and ocean currents. We, speaking as the collective average, are willing to let the world's climate and ecosystems be altered (the passive tense is important, see below) by our wastes from technological energy transformations that we require to live in the styles to which we have become accustomed or to which many still aspire. It isn't a physical response to the wastes themselves that causes disgust, for CO_2 is invisible and odorless. Nonetheless, a knowledge-grounded perception that something is wrong can still lead to feelings of disgust.

Are people going to take the time to be motivated by fear, anger, sadness, and disgust, say, about the fate of the polar bear, or will any such motivations be trumped by the ones attached to their fear of bear stock markets?

Results from an experiment in social psychology might have some relevance to the carbon challenge.[2] Its results were established before that time when the flow of future history becomes cloudy for me, except for seeing some snippets of the fates of our individual carbon atoms.

The experiment looked into a basic aspect of human morality, and uncovered a distinction between two kinds of caused death. You, if you are the subject in the experiment, are presented with two life-and-death circumstances and are asked what you would do. In both situations a runaway trolley is barreling down a track and headed to kill all five workers standing on the track just ahead, talking, and not seeing the onrushing train. In one situation, you can pull a lever to switch the trolley to a side track in the nick of time, but unfortunately one other worker you see standing on that side track, also oblivious to the situation, will be killed. In the second situation, you are on a bridge above the tracks and can push a fat man standing right next to you over the bridge and on to the track in front of the onrushing trolley. He is just the right size to stop the train. (Presumably you aren't, so there is no option for self-sacrifice.) The raw, objective math is the same in both cases: one dies to save five. In both situations you, as subject, are asked whether you would kill the one to save the five.

More people, on average, will approve the mechanical switch to the side track. Most will not opt for pushing the bystander to stop the train. This result has held up across different cultures. The social psychologists assert that through this and other experiments they are uncovering some deep, instinctual ethics. The difference in degree of physical contact with the one who gets killed (worker on side track or fat man on bridge) seems to connect to a felt difference in the degree of intention to kill. The case of the push is immediately felt as wrong. The case with the switched track is not as clear and can seem morally right to do.

The relevance of this experiment to the rise in CO_2 and global warming comes about because the pain that will be caused is relatively inadvertent, even though acknowledged. It might be difficult to achieve a gut level consensus of aversion toward the CO_2 rise of the kind that the social psychologists find evoked when people are faced with the prospect of pushing one onto a track to save five. The damage and harm from global warming can seem remote, definitely in space and certainly in time, more similar to diverting the train that kills one to save five. We vent our wastes that might inadvertently kill one (as an analogy to environmental damage, for instance) so that the rolling forward of the fossil-fuel-fired Global Industrial Growth Automaton can save five, with the analogy being the way that the current state of the GIGA is increasing the material well-being of most (although for many it is damn slow and they suffer). The harmful emissions are only a necessary though unfortunate by-product.

Imagine a world in which all countries had reached industrialization but without fossil fuels. (Say those fuels had not existed.) Perhaps the world energy source is nuclear or carbon-neutral biomass, either way there are no greenhouse problems. For one large, powerful, satellite-launching tropical nation, everything is just great, except that it's just too hot. The citizens feel that life would be so much better if their sweltering heat could be taken down a few notches. Forget air conditioners; they want the entire country cooler. So they hatch a plan to deploy a system of mirrors out in space to reflect some of the sun and gain relief. Neglecting the internal problems with their environmental impact statements there is one other pressing thorn in the plan. They cannot just fly the mirrors above their country, but must place them into high stable orbits that will affect and cool the entire Earth. In some regions outside their country to the farther north and south, growing seasons will be shortened, producing higher failure rates of crops. There would be immediate, international outrage of an instinctual nature. This fantasy would be more like the action of pushing the man onto the track.

This fantasy can be nudged a little closer to reality. Say things go badly for the United States with the future greenhouse climate, drying up the midwestern farm belts of corn and wheat. Imagine that NASA is then charged to design and deploy mirrors in space. But in this future, the crop yields of Canada already have been improved. Canada doesn't want space reflectors. It likes the new world. Will Canada, benefiting from the best climate in 100 years, shoot down the cargoes of mirrors sent up to into orbit? Or assume that climate change brings about more regular devastating floods to India and Bangladesh, who then hire China to launch the gargantuan array of mirrors. Perhaps Russia has gained an improved climate. Will Russia and Canada band together to take out the mirrors?

Who is going to be the one to decide to engineer climate? Even the decision to pull the CO_2 from the air will be an action that will affect all, for in the future the effects of the rise of CO_2 will be better known, and therefore the effects of reducing it will also be known. The beneficiaries are going to protest any pulling back. This is one main aspect of the carbon challenge, once the change becomes more obvious and underway along a path that is globally visible, like with the rolling back of the Arctic sea ice, which is already right in our faces.

Extreme climate events, when they start falling outside the normal regime of either frequency or magnitude, are going to be blamed on the altered climate. And therefore they are going to blamed on the people who have been the primary emitters and the beneficiaries of letting the fossil-fuel energy servants run as they vent the wastes out through the various piping outlets of those servants. International unrest will grow. Every climate disaster will be blamable. Popular conception will no longer pin sudden disasters or creeping change on "acts of God" but rather on "acts of Man."

Right now the tensions, already significant, are relatively mild relative to the potential for escalation. No one can be blamed for an actual intention to put out the gas wastes from the fossil-fuel energy servants to deliberately produce a certain role in climate which affects all. The intention is only to derive the material benefits from the energy servants. This lack of intention, as well as the lack of perceived immanent danger, is probably why more immediate, general revulsion stemming from fear, anger, sadness, and disgust is not there. But as the tide of information changes, as effects become more noticeable, then today's inadvertency of the emissions will start to be perceived more and more as a purposeful act to those who can claim that other options exist, in terms of energy systems and efforts toward conservation and efficiency. And there is starting to be a revulsion among developing countries who see that the industrialized countries have more responsibility because they have gotten wealthy because of their fossil-fuel energy servants.

During a meeting of the Group of Eight in June 2007, the president of Brazil said: "Everyone knows that the rich countries are responsible for 60 percent of the gas emissions, and therefore need to assume their responsibilities. . . . We don't accept the idea that the emerging nations are the ones who have to make sacrifices, because poverty itself is already a sacrifice."[3]

All human bodies make about the same amount of waste, in the form of CO_2 gas, and various compounds in their excreted bodily liquids and solids. Thus, when a public restroom smells bad because of overuse, no one person can be blamed. But equality is not the case in the gaseous excrement of the energy servants — the power plants, cars, factories, lights, air conditions, and so forth — that support the average citizens of various countries. A mountain range of inequity is at the core of the carbon challenge faced by people of the early twenty-first century.

The question stands: What is going to be the moral stance toward the pain that is caused, to people (and of course damage to ecosystems) that did not share much in the responsibility for the climate conditions that caused the pain? That is a really huge issue looming. And the United States and other high per capita emitters stand at the center of responsibility.

Whatever benefits there may be, even if they are judged beneficial for the world as an average (as noted, in terms of total photosynthesis, total rain, length of growing seasons), the world does not live as an average. The world does not live as an average per capita emitter of CO_2. Or per capita GWP. The world contains a great disparity in how people live, and there is great disparity in climate, and there is going to be a great disparity of greenhouse effects. Again, this is at the core of the challenge, all tangled up with inequality.[4]

The countries that emitted the gas disproportionately could say, in their own defense, that because they had access to the convenient forms of fossil-fuel energies in the twentieth century they were able to develop quickly and engage the scientists, engineers, and entrepreneurs who brought to market the technologies that have now become available to raise the wealth for anyone and anyplace on Earth. Therefore, the world as a whole will gain better living standards in the near future because of the use of coal, oil, and natural gas by the early emitters. Perhaps, in an alternative universe, the world could have reached its current state of development with a biomass-based energy structure. Or perhaps the world only got to where it is today because of the ease and beauties of the fossil fuels. Yes, the beauties. You can sometimes hear oil experts say "You know, you almost could not have asked God to have dialed us a more convenient fuel." If only it all could have been Saudi Arabian light crude, and more of it next time!

I exaggerate a little. But what is coming is heightened intensity around these kinds of issues. It will be too late to just turn down the emission spigots tomorrow at the snap of fingers. And nature has certain absorbing powers when running its natural course. You don't tell nature to just vacuum the CO_2 up when the best nature can do, as is evident from the fairly constant average airborne fraction over the previous 50 years, is about half of our waste emissions. Doing better would require more highly "engineered" ecosystems.

It would still be a complex matter, but certainly a simpler one, were the global greenhouse effect to be caused by everyone equally, so that everyone would have an equal piece of blame in the gamble and an equal stake in the results. But that is not the case.

Dave Exits the Biosphere

The deep ocean is not visited by carbon atoms as frequently as are the surface bowls of atmosphere or land plants. However, the long time a carbon atom spends in the deep ocean, when it does get there, is why the deep ocean is the bowl that claims the largest fraction of the atom's 100,000-year average lifetime in the biosphere.

When Dave entered the Antarctic Bottom Water, in 2012, he first traveled for about a century clockwise halfway around Antarctica. At the end of this partial circuit, up above in the air and across the lands the fossil-fuel era was coming to a close by the early years of the twenty-second century, whichever course the era took.

At that point, halfway around Antarctica, Dave was spun outward in one of the gigantic tongues of Antarctic Bottom Water that wedge underneath and travel northward into the Atlantic, Indian, and Pacific

Oceans. Dave happened to be in the Pacific tongue. He was slowly worked northward as the tongue crept along, staying deep due to its density and pushed by new deep water arriving from behind. And the carbon concentration of the water increased bit by bit. Organic debris from the well-lit surface layer miles above, though small in amount, did filter down to such deep depths. The debris was consumed as food by deep water bacteria, which ejected CO_2 as waste, which mostly became bicarbonate ions, joining Dave.

The Deep Water became less dense as it warmed, mixing somewhat with other waters above as it slowly traveled for more centuries, and Dave slowly drifted upward. Time passed in utter darkness. At a half-mile in depth, in the northern hemisphere, Dave hit the deep equator-bound loop of Intermediate Water and found himself going in reverse, back south but at a much shallower depth, and slowly heading up. Eventually he was in the Pacific Equatorial Upwelling, an ascending zone of water that surfaces in a belt near the equator. There in the tropics, a long stretch south of Hawaii, where he had been measured when in the air more than a millennium before, Dave came up into the surface mixed layer where the phytoplankton grow and the zooplankton swarm, where dolphins romp and swordfish hunt. He was edged up to the air-sea boundary. Suddenly, after 1,267 years in the ocean, he was—with a molecular ping—back in the air.

As it happened, Oiliver, Coalleen, and Icille were also in the air. Methaniel was still locked in carbonate precipitate, deep under the state of Kentucky. These three and Dave in individual CO_2 molecules now participated in absorbing and reemitting thermal infrared rays, until one by one, in only a few years, they were pushed or pulled along strands into other bowls of the biosphere.

In 3279, when Dave reemerged into the air after his long stay in the deep ocean, Earth's CO_2 level had been gradually decreasing for many centuries. As carbon-cycle scientists knew in the late twentieth century and in the early twenty-first, the excess levels of CO_2 from fossil-fuel combustion would eventually be reduced by slow mixing throughout the depths of the world ocean. But even after such a spell the level was substantially above what it had been before the start of the human-caused increase.

Dave cannot tell us what the peak level had been. We do know that it had been a long time since fossil fuels had been burned. There had simply not been enough of them to continue their use at the increasing rates of the late twentieth century and the early twenty-first century for more than about another century. So the fossil-fuel era ignited and flamed out rather quickly, a brief candle at the end of one phase of human history and the beginning of the next.

Now the millennia rolled on. And on. The atoms each followed individual paths. These, again, were roughly governed by the probabilities across the strands and bowls of the global carbon cycle, the same probabilities to which Dave had adhered during his first 32,000 years in the biosphere. No matter what happened to the human species, the carbon cycle carried on. Humans or their descendants had certainly not eliminated carbon from the atmosphere. CO_2 still blew forth from volcanoes. Carbonate ions were still being washed out and released from rocks such as limestone. The slow geological cycle of carbon would continue for as long as there were mountains and sea floors, given an Earth surface with rain, rock, rivers, and oceans. Yes, and air. And if there is life, so much the better. For then the adventures of the carbon atoms become that much more interesting, circuiting hundreds of times in and out of plants and algae and many other creatures.

But eventually the statistics took their toll, and the atoms found themselves in strands that led out of the biosphere.

The Ultimate Fates of Our Carbon Atoms

Coalleen went first, in what would have been the human year 56,374. The CO_2 level was down almost to what it had been in 1850. Coalleen was incorporated into the shell of a single-cell marine creature called a foraminiferan, which, upon dying, dropped down through the water and settled on the East Pacific Rise, an underwater mountain range off the western coast of South America, a place was sea floor was spreading. In the sediments, now covered by tons of other particles that had also fallen, she moved with the sea floor toward South America. She was on one of Earth's fastest-moving continental plates. And she was out of the active biosphere.

In year 83,945, Oiliver, in a microscopic particle of dead leaf matter, was washed down the Brahmaputra River across the region that had long ago been called Bangladesh, into the Ganges, then into that river's vast delta, where he was buried in mud. Usually he would have been eaten by sediment creatures and recycled but in this case he just happened to escape the clutches of burrowing creatures and the thick hosts of bacteria, and went out of the biosphere. He settled more and more deeply, destined eventually, millions of years later, to become a tiny bit of carbon in shale rock.

Icille simply went out in a piece of coral that was broken off by a storm in year 123,666. This coral was a descendant of species that humans had created through genetic modification near the end of the fossil-fuel era in their attempts to maintain the reefs against ocean acidification. Icille was carried out to sea and buried under a pile of fast-accumulating debris. She was out of the biosphere and she, too, would eventually become rock.

Dave had entered the biosphere before all the others, except Icille who had come with Dave. And he outlasted them all inside the biosphere, with a long stay granted by the luck of the dice thrown for all atoms that circulate so wondrously in and out of the places we see

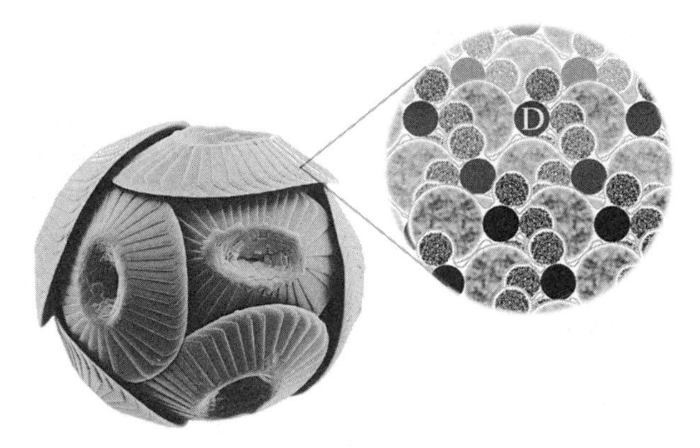

Figure 10.1 Dave the carbon atom in a crystal of calcium carbonate in the shell of a coccolithophore.

around us every day. Dave went out in the year 135,782, having been in the biosphere nearly 170,000 years. His exit from the great, complex stage was within the matrix of a shell of calcium carbonate (figure 10.1) made by a coccolithophore, a microscopic marine algae that precipitates its shell from the carbonate and calcium ions of seawater. It was just such a transfer from carbonate ion in water to calcium carbonate shell of a coccolithophore that had taken Dave out of the biosphere the previous time to a burial that had lasted 28 million years.

Similar to the exit path taken by Coalleen, the tiny shell fell downward when the coccolithophore died. The shell was above the Mid-Atlantic Ridge, the underwater mountain range that snakes like a spine down the middle of the Atlantic Ocean, with Europe and Africa to the east, with the Americas to the west. Settling near the top of the ridge, Dave was above the carbonate-dissolving deeper waters, and so was stable in the shell, which was covered year by year, in fact millions of years by millions of years, with a thick layer of more sediment wafting down from above.

Dave was slowly (at a pace about equal to the rate of fingernail growth) conveyed toward the Americas, as the Atlantic Ocean grew from the spreading new masses of sea floor formed by volcanism at the highest part of the ridge itself, now hundreds of miles behind Dave. At this time, 10 million years in the future, Coalleen, in the sediments of the fast-moving sea floor of the East Pacific, had been tunneled down under South America and heated to such an extent that she was "boiled off" into CO_2, which came up through a volcanic vent in the Andes. Thus she entered, once again, the biosphere.

At that 10-million-year point, Dave was not even close to reentering the biosphere, for the sea floor of the Atlantic Ocean had no active sites for geological recycling. It was not until the Atlantic Ocean stopped growing and the plate tectonic cycles reversed, tens of millions of years from now, that Dave even had a chance of reentering. It was then that the American continents started crawling at a few centimeters per year back toward Europe and Africa. In another 100 million years after that, the sediments in which Dave was buried underwater were scraped along the edge of a continent, like cleaning a bit of butter off the end of a knife. Over millions more years, he was lifted up and became a part of a crystalline limestone cliff on land, until the slow but steady erosion of rock layers exposed him to the sky. And then it rained.

Notes

Chapter 1

1. I direct the interested reader to some recent books that focus either more on climate or more on the economics and politics than my book does: Elizabeth Kolbert, *Field Notes from a Catastrophe: Man, Nature, and Climate Change* (Bloomsbury, 2005); Ross Gelbspan, *Boiling Point* (Basic Books, 2005); Timothy Flannery, *The Weather Makers: How Man Is Changing the Climate and What It Means for Life on Earth* (Atlantic Monthly Press, 2006); David Archer, *Global Warming: Understanding the Forecast* (Wiley, 2006); Bjorn Lomborg, *Cool It: The Skeptical Environmentalist's Guide to Global Warming* (Knopf, 2007); Ted Nordhaus and Michael Shellenberger, *Break Through: From the Death of Environmentalism to the Politics of Possibility* (Houghton Mifflin, 2007). In speaking about the environmental changes and challenges from the increasing concentrations of atmospheric CO_2, I use terms such as "greatest" and its synonyms (in the book's subtitle, for instance, and in the opening sentences of the preface and chapter 1). I do not mean to downplay other environmental problems, such as the global simplification of ecosystems and threats to water quality. That said, the CO_2 issue is unusual and "greatest" in its scale because it derives from the spreading out of a single chemical to fill the atmosphere. It affects the world everywhere, no matter the locations of its sources. Other environmental problems are global in a different way, coming from the summation of similar local impacts of economic systems. Their solutions can be separated to some extent. The quality of water in a particular river has to be solved within that

watershed, while improvements to a river on another continent will have to be focused there. CO_2 is globally greatest in the sense that its solution *must* be global, because that is the nature of the chemical and Earth's well-mixed air. Another global problem from a single, dispersed chemical is the depletion of stratospheric ozone due to CFCs, but that one, hopefully, is on the path to solution. For a cogent discussion about the issues of framing CO_2 with respect to other environmental concerns, see Eileen Crist, "Beyond the climate crisis: A critique of climate change discourse," *Telos* 141 (2007): 29–55.

2. All life and the three environmental matrices of atmosphere, soils, and oceans form a closely integrated network that can be called the "biosphere." Its upper boundary is clearly the top of the atmosphere. Its lower boundary is admittedly fuzzier. Groundwater reaches kilometers down into pores of rock, and bacteria have been found kilometers down too. But for practical purposes—in terms of pointing to the fact that creatures live within and exchange gases, liquids, and solids with the environmental matrices and for technical modeling of the impacts of organisms on the chemistry of the global environment on relatively short timescales—we can exclude from the definition of the biosphere the minerals in rocks underneath the soil, because the elements in those rocks have been out of active circulation for millions or hundreds of millions of years. Thus, in this book I use the term "biosphere" to refer to the integrated system of air, oceans, soils, and life.

3. To better see molecules, do a Web search aimed at images, using the name of the desired molecule and the term "space filling" or "3-D."

4. According to a personal communication from Ralph Keeling of the University of California at San Diego, asymmetric diatomic molecules (meaning that the two atoms are different, as in the case of carbon monoxide) also absorb infrared radiation, but not in wavelengths that are relevant to the atmosphere or to the climate.

5. Good discussions of this and the relative strengths of the air's suite of greenhouse gases can be found at www.realclimate.org. Also see the websites of NASA's Goddard Institute for Space Studies (www.giss.nasa.gov) and the Intergovernmental Panel on Climate Change (www.ipcc.ch).

6. More elaborate explanations can be found on various websites and in various textbooks on environmental science. One good textbook is Daniel Botkin and Edward Keller, *Environmental Science: Earth as a Living Planet* (sixth edition: Wiley, 2007). More complexity comes in than I have shown because all the infrared absorbing molecules at all altitudes of the atmosphere absorb and emit IR, and the IR that gets to space comes primarily from the upper atmosphere and very little directly from the surface.

Chapter 2

1. For a fuller discussion of the definition of the biosphere as an integrated system as air, ocean, soil, and life, see my book *Gaia's Body: Toward a Physiology of Earth* (MIT Press, 2003). That book also gives a more complete description of the steps of photosynthesis ("how Dave went from air to beer"), including the efficiencies of energy conversions in the various steps.

Chapter 3

1. CO_2 data from the website of the National Oceanographic and Atmospheric Administration (www.esrl.noaa.gov) are credited to Pieter Tans. The full graph of Mauna Loa's CO_2 is from www.esrl.noaa.gov. Data on CO_2 at other sites are from www.esrl.noaa.gov. Another good source for CO_2 data is the website of the Carbon Dioxide Information Analysis Center of the U.S. Department of Energy (http://cdiac.esd.ornl.gov).

2. Source: personal communication from Ralph Keeling.

3. Charles D. Keeling, "Rewards and penalties of monitoring the Earth," *Annual Review of Energy and Environment* 23 (1988): 25–82. This paper is marvelous both for its technical content and for providing an "inside look" at the process of science.

Chapter 4

1. This is the time of the paintings of the Chauvet cave in France.

2. For a fascinating analysis of the evolution of human cognition, see Steven Mithen, *The Prehistory of the Mind: The Cognitive Origins of Art, Religion and Science* (Thames and Hudson, 1999).

3. All units of tons in this book are metric tons, or 1,000 kilograms. One metric ton is approximately 10 percent larger than an American ton of 2,000 pounds. According to a personal communication from Robert Berner, a geochemist at Yale University, nature's annual rate of entry of about 400 million tons of carbon into the biosphere consists of about 100 million tons as CO_2 from volcanic outgassing, 250 million tons of carbon as carbonate ions dissolved by chemical weathering from calcium carbonate rocks, and 60 million tons as oxidized CO_2 from the weathering of rocks that contain organic carbon, such as shales. The combined CO_2 flux from volcanism and shale oxidation plays into determining the steady-state concentration of atmospheric CO_2 over geological timescales via the carbonate-silicate geochemical cycle. But the flux from carbonate dissolution does not. For more on these long-term processes of the global carbon cycle, see David Schwartzman, *Life, Temperature, and the Earth* (Columbia University Press, 2002).

4. Data on fossil-fuel fluxes are from the U.S. Department of Energy's Carbon Dioxide Information Analysis Center, supplemented for the years 2004-2006 by data from the *BP Statistical Review of World Energy June 2007.*

5. Vladimir I. Vernadsky, *Essays on Geochemistry and the Biosphere* (Synergetic Press, 2007).

Chapter 5

1. Data on historic emissions from changes in land use back to 1850 for various regions of the world, compiled by Richard "Skee" Houghton, can be obtained from the website of the U.S. Department of Energy's Carbon Dioxide Information Analysis Center.

2. A recent paper has set the average emissions from land use change as 1.5 billion tons of carbon annually, across recent decades, with an uncertainty of plus or minus a half billion tons (J. P. Canadell et al., "Contributions to accelerating atmospheric CO_2 growth from economic activity, carbon intensity, and

efficiency of natural sinks," *Proceedings of the National Academy of Sciences* 104 (2007): 18,866–18,870). I acknowledge conversations with Jon Foley about the numbers for carbon fluxes associated with land-use changes.

Chapter 6

1. Source of Law Dome ice-core CO_2 data: D. M. Etheridge et al., Law Dome Atmospheric CO_2 Data, IGBP PAGES/World Data Center for Paleoclimatology Data Contribution Series #2001-083, 2001 (available at www.noaa.gov).

2. Re Taylor Dome ice-core CO_2 data: Different citations are suggested by the NOAA for different time intervals. The data are available at www.noaa.gov.

3. The fact that the air's most common, light kind of carbon atoms (isotope carbon-12) in CO_2 molecules is increasing relative to the heavier kind (carbon 13) shows that the source of the rise is from biologically created matter. Fossil-fuel carbon does fit that bill. But so do modern plants and detrital carbon in the soil. The decrease (before atomic bomb tests) of the air's radioactive form of carbon (carbon-14) shows that source of the extra CO_2 comes from a form of carbon that is extremely ancient (known as the Seuss effect). Also of great interest is the fact that as fossil fuels are being burned, not only is CO_2 created but oxygen is consumed. We need not worry because of the great amount of oxygen in the atmosphere. Nonetheless, monitoring oxygen has proved invaluable for understanding the magnitudes of ecosystem fluxes to and from the atmosphere, both terrestrial and marine. One scientist in this line of study is none other than Dave Keeling's son, Ralph, based at the Scripps Institution for Oceanography in La Jolla, California. For a good technical but accessible paper on these issues, see F. Joos, "The atmospheric carbon dioxide perturbation," *Europhysics News* 27 (1996): 213–218.

4. Source of Vostok ice-core CO_2 data: J. R. Petit et al., "Vostok ice core data for 420,000 years," IGBP PAGES/World Data Center for Paleoclimatology Data Contribution Series #2001-076. NOAA/NGDC Paleoclimatology Program, Boulder. Again, the data are available at www.noaa.gov.

5. U. Siegenthaler et al., "Stable carbon cycle-climate relationship during the late Pleistocene," *Science* 310 (2005): 1313–1317.

Chapter 7

1. Fuller, who originally called these "energy slaves," made a trenchant point about how they raise our living standards in many of his writings. See, for example, *Critical Path* (St. Martin's Press, 1982).

2. The Netherlands Environmental Assessment Agency found that China's emissions surpassed those of the U.S. in 2006 when emissions from cement manufacture were added to emissions from fossil fuels. See www.mnp.nl/en.

3. Economic data (GDP and GWP) used in figure 7.3 are from the *CIA World Factbook 2007* (available at www.cia.gov). Energy data are from the *BP Statistical Review of World Energy June 2007* (available at www.bp.com). For CO_2 emissions, I used data from the Netherlands Environmental Assessment Agency (www.mnp.nl/en), which for the regions shown in the figure include fossil-fuel emissions only (not from cement manufacture, which I did not have for all cases; but see figure 4.4 for the historical global trend).

4. Data sources for figure 7.4: as for figure 7.3. (See preceding note.) ("EU-15" refers to the 15 nations in the European Union before its expansion in 2004.)

5. Data for figure 7.5 on gross world product over time are from the table for economic and population data compiled by German economist Angus Maddison, available at www.ggdc.net/maddison. I converted 1990 dollars from Maddison into 2005 dollars, and I supplemented his data (which terminate at 2003) with data from the *CIA World Factbook* for the individual years after correlating the two sources for a decade of overlap. They both use the standard economic concept of purchasing-power parity, though there is some contention on how that is calculated. My result of approximately 3% annual growth rate in real GWP is consistent with others (Marty Hoffert, personal communication). It is a standard result that is often cited.

6. See, for example, Vaclav Smil, *Energy at the Crossroads: Global Perspectives and Uncertainties* (new edition: MIT Press, 2005). See also various works on sustainability indicators by Herman Daly and others.

7. I am not immune to the drive for more. I think I'm fine, and then I read a news story about $5 million beach properties that the wealthy in the United States are buying along the coast of Costa Rica, and I find myself in a tizzy of

feeling deprived: why don't I have that, maybe this book will sell really well, and then all kinds of conflicts set in because I believe that not everyone can have such a lifestyle. But then (in answer to the conflict), if someone then why not me? It's called luxury envy. I consider it a sickness. But next the little inner voice says that I'm evolved to want and try and get, because those that did in our formative evolutionary past were reproductively successful and gave rise to others who also had those behaviors.

8. Economic data used in figure 7.6 are from Angus Maddison (see note 5). Energy data are from the U.S. Energy Information Administration (www.eia.doe.gov), particularly the Annual Report's chapter on International Energy. Another valuable source is the Annual Report of the International Energy Agency (www.iea.org). Data on CO_2 fluxes are from the Carbon Dioxide Information Analysis Center. A much-cited analysis of GWP, energy, and carbon emissions in terms of changing ratios is Martin I. Hoffert et al., "Energy implications of future stabilization of atmospheric CO_2 content," *Nature* 395 (1998): 881–884. This kind of analysis is also prominent in studies by the IPCC and by the International Institute for Applied Systems Analysis in Austria.

Chapter 8

1. Updates about these fascinating experiments can be found at cdiac.ornl.gov.

2. The paper is J. P. Canadell et al., "Contributions to accelerating atmospheric CO_2 growth from economic activity, carbon intensity, and efficiency of natural sinks," *Proceedings of the National Academy of Sciences* 104 (2007): 18,866–18,870. Canadell et al. found that the airborne fraction increased from about 0.40 to 0.45 over the nearly 50 years of the CO_2 record. Using a time series of the annual airborne fraction, I am able to reproduce their result. However, the statistical goodness of fit gives $r^2 = 0.003$, which is extremely low. One way to see the weakness is to perform the following data experiment: What if Keeling had started his measurements at Mauna Loa in 1965, rather than in 1959? We would not have the data for 1959–1964. Now, if I fit this "new" data set with a line, I obtain a *decreasing* airborne fraction over the time of record, of about the same magnitude as the increase claimed by Canadell et al. The result is weak if cutting a few years of data from one end of a rather lengthy time series changes the slope of the fitted line. In my opinion, it is the constancy of the airborne

fraction thus far over many decades that is the message from the record, though I agree that many lines of argument are correct in predicting that the airborne fraction will increase in the future.

3. N. Zheng, A. Mariotti, and P. Wetzel, "Terrestrial mechanisms of interannual CO_2 variability," *Global Biogeochemical Cycles* 19 (2005): GB1016. The strongest factor is the influence of the variability of rain during the El Niño southern oscillation on tropical vegetation and the balance of carbon fluxes in tropical ecosystems.

4. P. Friedlingstein, "Climate-carbon cycle feedback analysis: Results from the C4MIP model intercomparison," *Journal of Climate* 19 (2007): 3337–3353. I analyzed table 2 on page 3345 for averages and standard deviation, both with and without climate feedbacks.

5. I averaged the eleven scenarios in appendix II.1 of Houghton et al., *Climate Change 2001: The Scientific Basis* (Cambridge University Press, 2001). Emissions from land use relative to total emissions fell between 2% and 8% for most, the average being 5% of total cumulative emissions to 2050.

Chapter 9

1. Source: www.ipcc.ch.

2. A web search reveals numerous uses of this phrase by Hansen. Some of Hansen's popular writings are available at www.giss.nasa.gov.

3. K. Zweibel, J. Mason, and V. Fthenakis, "A solar grand plan," *Scientific American*, January 2008: 64–73.

4. M. I. Hoffert et al., "Advanced technology paths to global climate stability: Energy for a greenhouse planet," *Science* 298 (2002): 981–987.

5. P. Read, "Biosphere carbon stock management: Addressing the threat of abrupt climate change in the next few decades," *Climatic Change* 87 (2008): 343–346. If pristine ecosystems are converted to the production of biofuels from food crops, it can take decades or centuries to pay back the "carbon debt" incurred in the release of carbon dioxide from the conversion itself, from soil

and above ground biomass. See J. Fargione, J. Hill, D. Tilman, S. Polasky, and P. Hawthorne, "Land clearing and the biofuel carbon debt," *Science* 319 (2008): 1235–1238.

6. Jesse H. Ausubel, "Renewable and nuclear heresies," *International Journal of Nuclear Governance, Economy and Ecology* 1 (2007): 229–243; James Lovelock, *The Revenge of Gaia: Earth's Climate Crisis and the Fate of Humanity* (Perseus Books, 2006).

7. There are useful websites on carbon sequestration and on other topics discussed in this chapter, including geoengineering, ocean acidification, nuclear fusion, peak oil, bioenergy, and biofuels.

8. Q. Schiermeier, "Europe to capture carbon," Nature 451 (2008): 232.

9. See, e.g., proposals by Klaus Lackner of Columbia University. A different candidate is a proposal by Greg Rau and Ken Caldeira to sequester CO_2 in the ocean by "accelerated weathering of limestone," a more environmentally benign means than direct injection.

10. O. Hoegh-Guldberg et al., "Coral reefs under rapid climate change and ocean acidification," *Science* 318 (2007): 1735–1742.

11. S. Pacala and R. Socolow, "Stabilization wedges: Solving the climate problem for the next 50 years with current technologies," *Science* 305 (2004): 968–972.

12. For an overview of the global energy situation that emphasizes the pressing need for fundamental research, see Hoffert et al., "Advanced technology paths to global climate stability." Another strong argument for a giant increase in government-funded energy research has been made by Ted Nordhaus and Michael Shellenberger. (See their 2007 book *Break Through*.) Hoffert's appearance on *The News Hour* is worth viewing on youtube.

13. Cartoon by Dwayne Powell, *Raleigh News and Observer*, 1989.

14. There are different ways of approaching the issue of costs. For some people, the chances of great disruptions to the biosphere and thus human well-being justify spending anything now to prevent the possibility. For others, they want

to see a bottom-line tradeoff between ways to spend money now and damages in the future from climate change. This is related to the issue of economic discounting, because it is argued that money not spent now to prevent climate change but spent for other purposes of general economic development will make the world wealthier in the future and therefore better able to afford expenditures later, at such time if and when climate change becomes a problem. Two major debaters on the topic of how to discount the future are the economists William Nordhaus and Nicholas Stern—see, e.g., *Nature* 317 (2007): 201–204.

15. M. Raupach et al., "Global and regional drivers of accelerating CO_2 emissions," *Proceedings of the National Academy of Sciences* 104 (2007): 10288-10293.

16. The term is from the title of a book on the consequences of peak oil production: James H. Kunstler, *The Long Emergency: Surviving the End of Oil, Climate Change, and Other Converging Catastrophes of the Twenty-First Century* (Grove Press, 2006).

17. I analyzed data in the appendix of J. T. Houghton et al., *Climate Change 2001: The Scientific Basis* (Cambridge University Press. 2001), which gave results for CO_2 levels from various "SRES" emissions scenarios applied to two carbon-cycle models. I used the following airborne fractions for my three scenarios, based on a fit between resulting overall airborne fraction for each scenario and its total emissions: 0.45 percent for the Constant Fossil Fuel, 49 percent for Central Trend, and 53 percent for Frozen-2010 Technology. These numbers cause a larger spread in the CO_2 results than with a single constant airborne fraction, but the conclusions are still the same because the spread either way is rather small.

18. I used the equation with four exponential decay terms in J. Hansen et al., "Dangerous human-made interference with climate: A GISS modelE study," *Atmospheric Chemistry and Physics* 7 (2007): 2287–2312. They cite the equation from F. Joos et al., "An efficient and accurate representation of complex oceanic and biospheric models of anthropogenic carbon uptake," *Tellus* 48B (1996): 387-417. My results were within a few ppm of the results I obtained with an airborne fraction model.

19. If the climate sensitivity for CO_2 doubling is lower than 3°C (say, 2°C) or higher than 3°C (say, 4°C), then all the values on the scale in the figure are

simply changed proportionately. So we can compare scenarios in terms of relative perturbations independent of the climate sensitivity assumed.

20. From the Intergovernmental Panel on Climate Change Fourth Assessment Report, *Climate Change 2007: Synthesis Report, Summary for Policymakers*, downloaded from the IPCC, January 27, 2008: "Contraction of the Greenland ice sheet is projected to continue to contribute to sea level rise after 2100. Current models suggest virtually complete elimination of the Greenland ice sheet and a resulting contribution to sea level rise of about 7 m if global average warming were sustained for millennia in excess of 1.9 to 4.6°C relative to pre-industrial values. The corresponding future temperatures in Greenland are comparable to those inferred for the last interglacial period 125,000 years ago, when palaeoclimatic information suggests reductions of polar land ice extent and 4 to 6 m of sea level rise." The climate scientist James Hansen warned that, based on nonlinear dynamics of ice sheets that are not well understood and on the paleoclimatic record, we could be committing ourselves to substantial changes in sea level that could occur on time scales of centuries not millennia. See, e.g., J. Hansen, "A slippery slope: How much global warming constitutes 'dangerous anthropogenic interference'?" *Climatic Change* 68 (2005): 269–279.

21. Quirin Schiermeier, "Europe spells out action plan for emissions targets," *Nature* 451 (2008): 504–505.

22. For this disturbing but perhaps not surprising finding, see D. Rivlin, "The millionaires who don't feel rich," *New York Times*, August 5, 2007.

23. In "A solar grand plan," Zweibel et al. estimate the costs at $420 billion in subsidies from 2011 to 2050. The full costs of the Iraq War, including veterans' benefits in the future, have been projected to be as high as $2 trillion.

24. Consider the overall carbon efficiency of economic productivity as the ratio of GDP to CO_2 emissions. The US carbon efficiency has improved by a cumulative factor of nearly 2 over about the last 30 years. According to my calculations, for the US to reduce its carbon emissions by half by 2050 while growing economically by 2–3% per year, it will have to annually improve the carbon efficiency of economic productivity at the rate of 1.75–2.25 times the annual average rate of improvement over the last 30 years. In my opinion this

is possible because up to now the improvements have been achieved without institutionalizing them as a comprehensive goal.

Chapter 10

1. The motivation to take action or change behavior to counter a risk depends on the degree of personal, visceral experience with the risk. See E. U. Weber, "Experience-based and description-based perceptions of long-term risk: Why global warming does not scare us (yet)," *Climatic Change* 77 (2006): 103–120.

2. See, e.g., Marc D. Hauser, *Moral Minds: The Nature of Right and Wrong* (HarperCollins, 2006).

3. L. Rohter, "Brazil, alarmed, reconsiders policy on climate change," *New York Times*, July 31, 2007.

4. For an eloquent analysis of the issue of global carbon inequity, see Tom Athanasiou and Paul Baer, *Dead Heat: Global Justice and Global Warming* (Seven Stories Press, 2002). Also visit the website www.ecoequity.org.

Index

DISCARD